PAIN PALLIATION OF BONE METASTASES: PRODUCTION, QUALITY CONTROL AND DOSIMETRY OF RADIOPHARMACEUTICALS

The following States are Members of the International Atomic Energy Agency:

AFGHANISTAN
ALBANIA
ALGERIA
ANGOLA
ANTIGUA AND BARBUDA
ARGENTINA
ARMENIA
AUSTRALIA
AUSTRIA
AZERBAIJAN
BAHAMAS
BAHRAIN
BANGLADESH
BARBADOS
BELARUS
BELGIUM
BELIZE
BENIN
BOLIVIA, PLURINATIONAL STATE OF
BOSNIA AND HERZEGOVINA
BOTSWANA
BRAZIL
BRUNEI DARUSSALAM
BULGARIA
BURKINA FASO
BURUNDI
CAMBODIA
CAMEROON
CANADA
CENTRAL AFRICAN REPUBLIC
CHAD
CHILE
CHINA
COLOMBIA
COMOROS
CONGO
COSTA RICA
CÔTE D'IVOIRE
CROATIA
CUBA
CYPRUS
CZECH REPUBLIC
DEMOCRATIC REPUBLIC OF THE CONGO
DENMARK
DJIBOUTI
DOMINICA
DOMINICAN REPUBLIC
ECUADOR
EGYPT
EL SALVADOR
ERITREA
ESTONIA
ESWATINI
ETHIOPIA
FIJI
FINLAND
FRANCE
GABON
GAMBIA
GEORGIA
GERMANY
GHANA
GREECE
GRENADA
GUATEMALA
GUYANA
HAITI
HOLY SEE
HONDURAS
HUNGARY
ICELAND
INDIA
INDONESIA
IRAN, ISLAMIC REPUBLIC OF
IRAQ
IRELAND
ISRAEL
ITALY
JAMAICA
JAPAN
JORDAN
KAZAKHSTAN
KENYA
KOREA, REPUBLIC OF
KUWAIT
KYRGYZSTAN
LAO PEOPLE'S DEMOCRATIC REPUBLIC
LATVIA
LEBANON
LESOTHO
LIBERIA
LIBYA
LIECHTENSTEIN
LITHUANIA
LUXEMBOURG
MADAGASCAR
MALAWI
MALAYSIA
MALI
MALTA
MARSHALL ISLANDS
MAURITANIA
MAURITIUS
MEXICO
MONACO
MONGOLIA
MONTENEGRO
MOROCCO
MOZAMBIQUE
MYANMAR
NAMIBIA
NEPAL
NETHERLANDS
NEW ZEALAND
NICARAGUA
NIGER
NIGERIA
NORTH MACEDONIA
NORWAY
OMAN
PAKISTAN
PALAU
PANAMA
PAPUA NEW GUINEA
PARAGUAY
PERU
PHILIPPINES
POLAND
PORTUGAL
QATAR
REPUBLIC OF MOLDOVA
ROMANIA
RUSSIAN FEDERATION
RWANDA
SAINT KITTS AND NEVIS
SAINT LUCIA
SAINT VINCENT AND THE GRENADINES
SAMOA
SAN MARINO
SAUDI ARABIA
SENEGAL
SERBIA
SEYCHELLES
SIERRA LEONE
SINGAPORE
SLOVAKIA
SLOVENIA
SOUTH AFRICA
SPAIN
SRI LANKA
SUDAN
SWEDEN
SWITZERLAND
SYRIAN ARAB REPUBLIC
TAJIKISTAN
THAILAND
TOGO
TONGA
TRINIDAD AND TOBAGO
TUNISIA
TÜRKİYE
TURKMENISTAN
UGANDA
UKRAINE
UNITED ARAB EMIRATES
UNITED KINGDOM OF GREAT BRITAIN AND NORTHERN IRELAND
UNITED REPUBLIC OF TANZANIA
UNITED STATES OF AMERICA
URUGUAY
UZBEKISTAN
VANUATU
VENEZUELA, BOLIVARIAN REPUBLIC OF
VIET NAM
YEMEN
ZAMBIA
ZIMBABWE

The Agency's Statute was approved on 23 October 1956 by the Conference on the Statute of the IAEA held at United Nations Headquarters, New York; it entered into force on 29 July 1957. The Headquarters of the Agency are situated in Vienna. Its principal objective is "to accelerate and enlarge the contribution of atomic energy to peace, health and prosperity throughout the world".

IAEA Radioisotopes and Radiopharmaceuticals Series No. 9

PAIN PALLIATION OF BONE METASTASES: PRODUCTION, QUALITY CONTROL AND DOSIMETRY OF RADIOPHARMACEUTICALS

INTERNATIONAL ATOMIC ENERGY AGENCY
VIENNA, 2023

COPYRIGHT NOTICE

Marketing and Sales Unit, Publishing Section
International Atomic Energy Agency
Vienna International Centre
PO Box 100
1400 Vienna, Austria
fax: +43 1 26007 22529
tel.: +43 1 2600 22417
email: sales.publications@iaea.org
www.iaea.org/publications

Printed by the IAEA in Austria
July 2023
STI/PUB/2042

IAEA Library Cataloguing in Publication Data

Names: International Atomic Energy Agency.
Title: Pain palliation of bone metastases : production, quality control and dosimetry of radiopharmaceuticals / International Atomic Energy Agency.
Description: Vienna : International Atomic Energy Agency, 2023. | Series: IAEA radioisotopes and radiopharmaceuticals series, ISSN 2077-6462 ; no. 9 | Includes bibliographical references.
Identifiers: IAEAL 23-01583 | ISBN 978–92–0–150022–9 (paperback : alk. paper) | ISBN 978–92–0–150122–6 (pdf) | ISBN 978–92–0–150222–3 (epub)
Subjects: LCSH: Radiopharmaceuticals. | Radiopharmaceuticals — Quality control. | Nuclear medicine.
Classification: UDC 615.849.6 | STI/PUB/2042

FOREWORD

Metastases are the major complication of cancer, and the most common cancers, such as those of the breast, prostate and lung, are more vulnerable to bone metastasis. Excruciating pain is the most fearsome condition of advanced bone metastasis, along with other conditions such as pathological fractures and spinal cord compression that result in significant morbidities and affect quality of life.

Multimodality treatment options are usually used to manage bone metastasis and provide bone pain palliation. Radiation therapy and pharmacological patient-centric management based on the analgesic ladder are the most common of these. Radionuclide therapy using bone targeting radiopharmaceuticals is an effective palliative care option to reduce pain associated with advanced disease. It involves the administration of a radiopharmaceutical with a particulate radiation emitting radionuclide. The treatment is often performed as an outpatient procedure and the patient can be discharged after a few hours in the nuclear medicine department. The pain relief starts within a few days and continues for several weeks, depending on the radiopharmaceutical being used. The therapy is safe, with no major significant side effects.

Bone pain palliation therapy using radiopharmaceuticals has been practised for several decades, but its benefits have been unavailable to a large number of patients, mainly owing to the limited availability and cost of the radiopharmaceuticals. Over the last 20 years significant work has been made on developing radiopharmaceuticals using radionuclides that are now more widely available. The IAEA has also played a major role in promoting research as well as clinical use of the products in Member States through coordinated research projects.

This publication is the outcome of a consultants' meeting, at which an international team of experts identified various aspects of the development of bone seeking radiopharmaceuticals, including radionuclide production, the development of new target seeking molecules, in vitro and in vivo evaluation of the new products, and estimation of the absorbed doses delivered. This publication compiles the above information on the development of radiopharmaceuticals and is expected to benefit scientists and professionals working on the development of similar products in different Member States.

The IAEA thanks the experts who contributed to this publication. The valuable assistance of M.R.A. Pillai, Molecular Group of Companies, Cochin, India, for compiling and editing this publication is gratefully acknowledged. The IAEA officer responsible for this publication was A. Korde of the Division of Physical and Chemical Sciences.

CONTENTS

1. INTRODUCTION

1.1. BACKGROUND

According to the World Health Organization's statistics, approximately 1 in 6 deaths globally are due to cancer [1]. Surgery, chemotherapy, targeted therapy and radiation therapy are the major treatments for addressing cancer. Diagnostic nuclear medicine plays a pivotal role in cancer management, as single photon emission computed tomography (SPECT)[1] and positron emission tomography (PET) imaging are clinically significant for detection, staging and screening of patients for different types of tumours and for post-therapy follow-up. Therapeutic nuclear medicine also plays a niche role in the management of some types of cancers, such as that of the thyroid, neuroendocrine tumours and prostate cancer. It is very likely that the current momentum for developing target seeking radiopharmaceuticals will continue and useful radiopharmaceuticals for the treatment of other cancers will also be developed in the near future [2, 3].

Patients in advanced stages of cancer often suffer from metastasis in bone, which is most common in prostate, breast and lung cancers, which account for nearly 45% of total cancers. Studies report that nearly 30% of patients with advanced stage lung, bladder and thyroid cancer and nearly 70% of patients with advanced stage prostate and breast cancer develop metastasis in bone [4]. Severe pain, fractures and spinal cord compression are some of the complications associated with bone metastases. Impaired mobility and sleep disturbance, in addition to unbearable constant nagging pain or episodes of acute shooting types of pain, result in severely compromised quality of life for patients. Hence, the management of bone metastases is a major issue in all clinical set-ups.

There are several options for the management of patients with metastatic bone pain. These include chemotherapy, hormone therapy, radiotherapy, bisphosphonate therapy and the use of other analgesics. Analgesic treatment starts with non-steroidal anti-inflammatory drugs (NSAIDs) and usually narcotic analgesics become essential as the intensity of pain increases; stronger opioids are gradually employed in spite of their known common side effects of sedation and addiction. With localized disease there is the option of using external beam radiotherapy as a treatment modality [4, 5]. The use of bone seeking radiopharmaceuticals with radionuclides emitting high linear energy transfer (LET) particulate radiation is a viable option for the management of widespread metastases [4–6].

---

[1] A list of the abbreviations used in the text is given at the end of the publication.

The IAEA has given special emphasis to the management of bone metastases by supporting research on the development of radiopharmaceuticals for targeted therapy as well as clinical trials of some of the radiopharmaceuticals developed through coordinated research projects (CRPs) [7–9].

1.2. OBJECTIVE

In 2007, the IAEA published IAEA-TECDOC-1549, Criteria for Palliation of Bone Metastases — Clinical Applications, which discussed in detail bone pain palliation using external beam radiation therapy (EBRT) and radionuclide therapy [4]. This publication provided relevant details about patient selection and preparation, recommended dose and efficacy of treatment with the radiopharmaceuticals that were available and in use at that time. There have been significant developments in the field regarding the development of new radiopharmaceuticals for bone pain palliation since the publication of IAEA-TECDOC-1549 [4]. These include the use of new radionuclides as well as better targeting mechanisms. The objective of the present publication is to provide details regarding the production of the radionuclides and radiopharmaceuticals used for bone metastases. The information provided in this publication will be useful to professionals in the field engaged in the development and deployment of bone seeking radiopharmaceuticals for the benefit of a larger number of patients.

1.3. SCOPE

This publication provides details on the development and production of radiopharmaceuticals for bone pain palliation. The availability of radionuclides is an important consideration in the development of radiopharmaceuticals. A number of radionuclides are used for the preparation of bone pain palliation agents and the production and purification of these radionuclides are detailed in this publication. The development of radiopharmaceuticals is an involved science requiring intricate understanding of chemistry, biology and medical needs. This publication details the development of bone pain palliation radiopharmaceuticals with different radionuclides and carrier molecules. It also provides the manufacturing formulae for the preparation of the currently used bone pain palliation radiopharmaceuticals.

1.4. STRUCTURE

This publication is organized into different sections. Section 2 provides basic details about bone metastasis and the different options used to manage pain. The development of radiopharmaceuticals for bone pain palliation involves the selection of a suitable target seeking molecule, the selection of a radionuclide, the preparation and evaluation of the radiotracers to determine their suitability as bone targeting agents, and the use of clinical trials using lead molecules. These aspects of the design of radiopharmaceuticals for bone pain palliation are discussed in Section 3. Production of radionuclides is important for the sustained availability of radiopharmaceuticals and Section 4 discusses the production of radionuclides used for bone pain palliation in detail. The radiopharmaceuticals developed ought to undergo a battery of in vitro and in vivo evaluations, and these are discussed in Section 5.

The efficacy of targeted therapy depends on the radiation dose that is delivered to the target while sparing healthy tissues. Detailed studies to estimate the dose delivered by a radiopharmaceutical to the target and non-target organs are a requirement prior to applying the agent for clinical trial. Section 6 provides details about radiation dosimetry studies. A number of bone pain palliation radiopharmaceuticals are currently available, with varying extents of application, and these products are described in Section 7. Target seeking radiopharmaceuticals have a role in bone pain palliation, as the uptake of these tracers in cancer cells is based on the biological properties of the tumour rather than the uptake mechanism, such as acting as a calcium mimic, or uptake in calcium hydroxyl apatite in the bone compartment. Section 8 provides details about these specific tumour targeted radiopharmaceuticals that accumulate in metastatic bone sites. The latest updates on these types of products are also covered in brief in this section.

Radiopharmaceuticals are generally obtained as finished products in doses that are ready for patient administration. However, these products are also formulated at hospital radiopharmacies by using freeze-dried kits with a radionuclide solution that has been sourced independently. Of late, radiopharmacies have also been formulating some of the target seeking radiopharmaceuticals, especially ^{177}Lu and ^{225}Ac based ones, as per a specific patient's needs. Details concerning the production of and quality control for the freeze-dried kits, as well as target seeking radiopharmaceuticals, are provided in Section 9.

2. BONE METASTASIS AND MANAGEMENT

2.1. BONE

Bone is an important component of the skeletal system and it has multiple functions, such as supporting the body structurally, protecting vital organs and allowing body movements. Bone marrow, where new blood cells are formed, is held within the bone. Bone also acts as a storage site for minerals, of which calcium and phosphorus are the most important. Both these elements are metabolically important and they are present in the bone in a similar chemical form to the synthetic mineral calcium hydroxyapatite (HA) ($Ca_{10}(PO_4)_6(OH)_2$) [10].

Bone undergoes continuous remodelling by osteoblastic activity in which new bone tissues are formed. In order to balance this bone formation, part of the bone is also destroyed through osteoclastic activity [11]. These two processes are well regulated within the bone to maintain bone structure and quickly repair bone damage.

2.2. BONE METASTASIS

As with most other organs, bone cells are also susceptible to cancer formation, although primary bone cancers are relatively rare. Primary bone cancers account for less than one per cent of all cancers detected. Osteosarcoma, chondrosarcoma, Ewing's sarcoma and chordoma are the reported forms of primary bone cancer [12]. However, bone is the preferred secondary growth site for some common cancers, including breast, prostate and lung, which show a tendency to metastasize to the bone. These secondary cancers that spread and grow in the bone are called metastatic bone cancers.

A primary cancer releases cancer cells and these cells travel to other parts of the body through either the blood or the lymphatic fluid. Most of these cancer cells will be killed by the immune system because they are not recognized as normal cells. However, some cancer cells can escape attack from the immune system and reach other organs and start to grow, thereby developing secondary cancers in other body parts. Bone tissue possesses a unique microenvironment that facilitates the homing and growth of cancer cells. Bone marrow, being rich in nutrients, provides the right environment for the homing and multiplication of cancer cells. Several biochemicals and proteins secreted by cancer cells also interfere with the process of bone formation as well as bone resorption [13]. After the liver and lungs, bone is the most susceptible site for cancer metastasis.

Bone metastasis induces changes in the bone structure and damages portions of the bone. The formation of holes in the bone is common. These holes in the bone are called osteolytic lesions. It is well known that normal bone remodelling is regulated by a balance between bone formation mediated by osteoblasts and bone resorption by osteoclasts [11]. This balance is disrupted not only by various metabolic bone diseases such as osteoporosis, Paget's disease, tumour associated hypercalcaemia and osteolysis, but also by bone metastases [14, 15].

A large percentage of patients in the advanced stages of breast, prostate, gastrointestinal and lung cancers are likely to develop bone metastases [16]. These metastatic bone lesions have a huge impact on the quality of life experienced by patients, as these lesions cause various complications, such as severe pain, pathological fractures, hypercalcaemia and spinal cord compression. Hence, the management of metastatic bone pain is an important aspect of cancer care [17, 18].

2.3. EFFECT OF BONE METASTASIS

Bone metastasis leads to chronic pain. Pain can be due to bone destruction caused by substances such as tumour necrosis factor alpha and other cytokines produced by tumour and inflammatory cells. Nerve compression and invasion into adjacent tissues can cause localized pain. Increasing tumour sizes also exert pressure on adjacent tissues and cause deformation of the bone [19–21]. Pain perception is a complex process involving nerves, spinal cord and brain. Metastasized bone cancer cells are known to stimulate local inflammatory mediators, which causes an acidic environment, stimulate the peripheral nerve endings in bone and activate sensory neurons. Severity of pain is related to neurochemical changes that occur at the dorsal root ganglia region of the spinal cord [21, 22]. Weak bones occasionally result in pathological fractures and breakages, causing severe pain.

2.4. DIAGNOSIS OF BONE METASTASIS

Imaging techniques such as computed tomography (CT), magnetic resonance imaging (MRI), SPECT and PET are useful to evaluate the extent of bone metastasis [23–25]. Tissue biopsy is usually performed as a confirmatory test. SPECT imaging using a radiopharmaceutical called ^{99m}Tc–MDP (methylene diphosphonate) is one of the common tests in a nuclear medicine department to detect bone metastasis. Being a radiopharmaceutical containing phosphonate groups, ^{99m}Tc–MDP participates in the phosphorus related metabolic activity taking place within the bone. A whole body scan is performed 2–4 h after the injection of 740–1110 MBq (20–30 mCi) of ^{99m}Tc–MDP. Figure 1 (left) shows a SPECT image using ^{99m}Tc–MDP in a patient with multiple bone metastases. Bone scans with ^{99m}Tc–MDP are also used to distinguish metabolic disorders of the bone from diseases other than cancer. These include infection, arthritis and bone fractures.

A PET–CT scan using sodium fluoride (Na^{18}F) is also used to determine the extent of bone metastases. First, 185–370 MBq (5–10 mCi) of activity is injected into the patient. Then both CT and PET images are taken 1–2 h after injection of the radiopharmaceutical. The bone images from sodium fluoride are superior to those from ^{99m}Tc–MDP. Sodium fluoride is both sensitive and specific for the detection of lytic and sclerotic malignant lesions and can also differentiate between benign and malignant lesions. PET–CT images derived using sodium fluoride also allow quantification of the disease by measuring standard uptake values and this can be used to assess the response to treatment as well as to identify the progression of disease relief [26]. Figure 1 (middle) is a sodium fluoride PET–CT image of a breast cancer patient with extensive bone metastases.

Bone metastasis cells will overexpress the same types of oncoproteins, receptors and enzymes that are overexpressed by the primary cancer. For example, if a prostate cancer metastasizes in bone, the bone cancer cells developed will have the same features as those of prostate cancer cells. Thus, a targeting agent used for imaging a primary cancer can be used for imaging bone metastasis. Figure 1 (right) shows a PET–CT image of a patient suffering from metastatic prostate cancer. The radiopharmaceutical used is ^{68}Ga–PSMA-11, which is a targeting agent for the prostate specific membrane antigen (PSMA), which is overexpressed in prostate cancer cells [27].

2.5. MANAGEMENT OF BONE METASTASIS WITH THERAPEUTIC DRUGS

The major goal of bone metastasis treatment is pain control and quality of life improvement. Treatment leading to a reduction in the size of the tumour is also highly desirable. Treatment options are based on multiple factors, such as age, life expectancy, pain score and quality of life, as well as the overall status of clinical disease. There are several approaches for managing metastatic bone pain and these are discussed in the following sections [21].

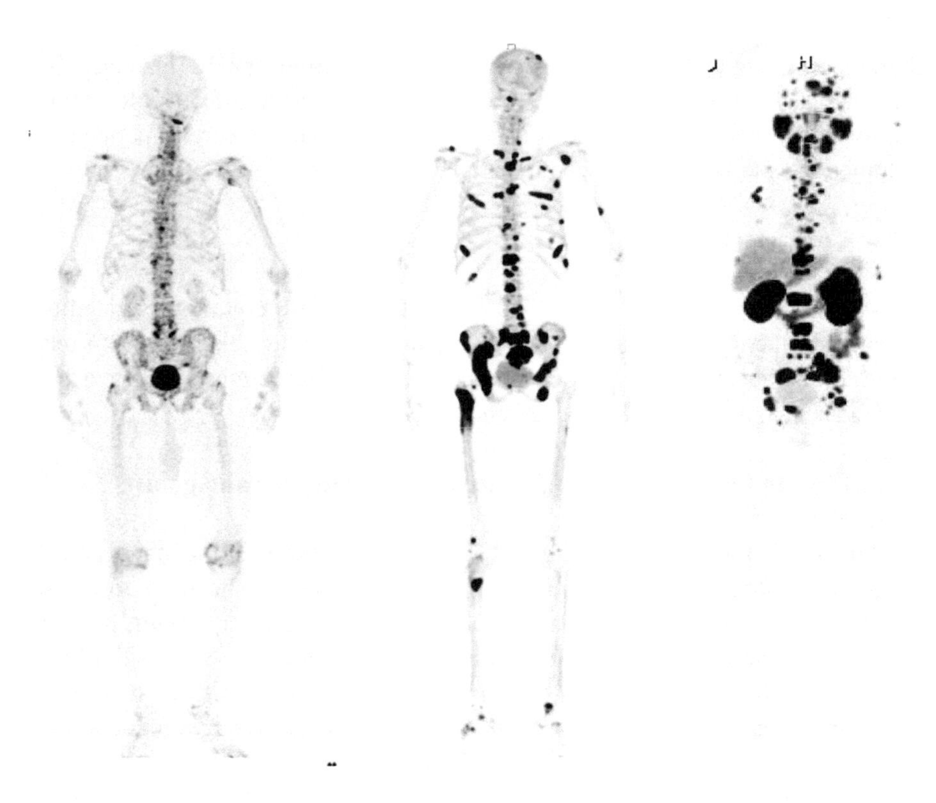

FIG. 1. SPECT scan using ^{99m}Tc–MDP of a prostate cancer patient with multiple bone metastases (left). Sodium fluoride (Na^{18}F) PET–CT imaging of a breast cancer patient with extensive bone metastasis (middle). PET–CT imaging with ^{68}Ga–PSMA-11 of a prostate cancer patient with extensive metastasis (right) (courtesy of M.R.A. Pillai, Molecular Group of Companies).

2.5.1. Non-steroidal anti-inflammatory drugs

NSAIDs are commonly used analgesics. Due to the anti-inflammatory effect of these drugs, they are also useful for treating pain caused by inflammation arising from cancer tissue invasion and destruction [28]. The short duration of their therapeutic effect limits the use of these drugs for relieving severe, constant pain conditions such as metastatic bone pain [21].

2.5.2. Opioids

Opioid drugs or narcotic analgesics are used in nearly 80% of terminally ill cancer patients due to their long lasting analgesic potential. Physiological dependence, tolerance that decreases the analgesic effect of the dose and respiratory depression associated with the use of these drugs limits their use, although other side effects such as nausea and vomiting can be managed by simultaneously administering other drugs. Opioids are very effective for acute pain, but long term use of opioid drugs also worsens pain in certain patients, a phenomenon called opioid induced hyperalgesia [21].

2.5.3. Corticosteroids

Anti-inflammatory drugs such as corticosteroids are commonly added as adjuvants for the management of bone pain due to metastatic disease [29]. Long term use of corticosteroids is associated with immune suppression, hypertension, hyperglycaemia, gastric ulcers and psychosis [21]

2.5.4. Growth factors, signalling molecules and receptor antagonists

Growth factors and signalling molecules are also used for the management of metastatic bone pain [30]. For example, osteoprotegerin is used to arrest osteoclastic activity. Also known as osteoclastogenesis inhibitory factor, it binds to the receptor activator of nuclear factor kappa B ligand (RANKL) with high affinity. This inhibits bone destruction but activates the apoptosis of osteoclasts. This apoptosis causes a reduction in the amount of bone damage; reduction in pathological fractures helps in pain reduction [21].

Endothelin groups of peptides with vasoconstriction activity are produced by endothelial cells. Endothelin and its receptors trigger several pathways of signal transduction, hence their role in various biochemical processes associated with propagation of cancer is now well studied. One of the endothelin receptor subtypes is found to be associated with bone metastasis occurring due to prostate and breast cancer. The drugs that selectively inhibit this protein are thought to play an important role in controlling disease spread to bone [31].

2.5.5. Bisphosphonates

Bisphosphonates, or diphosphonates, are molecules that have two phosphonate (PO(OH)2) groups in their structures. Bisphosphonates are known to inhibit the loss of bone density and are commonly used for the treatment of osteoporosis and other similar diseases [32]. It is well documented in the literature that bone undergoes constant remodelling by maintaining a fine balance between osteoblasts, which form bony tissues, and osteoclasts, which destroy such tissues. Bisphosphonates are known to prevent osteoclastic activities, thereby reducing bone loss [21]. This class of molecules is also used for the treatment of skeletal metastases, as these chemical compounds have the ability to reduce the acidic microenvironment of bone tumour tissues. The reduction of the acidic microenvironment reduces the dissolution of the bone and consequently reduces the pain associated with metastatic cancers [33]. Bisphosphonate drugs, such as alendronate, zoledronate, risedronate, medronate, oxydronate and others, have been proven to be useful when radiation therapy and analgesic medicines become ineffective in controlling bone pain [34].

2.5.6. Denosumab

Increased expression of RANKL is associated with metastatic bone cells, and leads to increased osteoclast activity, bone destruction and bone pain [35]. Denosumab, a fully humanized monoclonal antibody that specifically inhibits RANKL, is an approved drug for the prevention of skeletal related pain in patients with bone metastases from solid tumours and the treatment of osteoporotic fractures in postmenopausal women [36].

2.6. MANAGEMENT OF BONE METASTASIS WITH RADIATION

2.6.1. General principles for the use of ionizing radiation for cancer management

High energy ionizing radiation transfers its energy to the medium through which it passes. The human body is made up of about 36 trillion cells. Ionizing radiation interacts with cells and transfers its energy to the cells, initiating chemical reactions within them.

Deoxyribonucleic acid (DNA) molecules, which are a constituent of the nucleus of living cells, will be affected, depending on how much energy is delivered to the cells and imparted to the DNA. The mechanistic vision of DNA damage mostly comprises ballistics. There can be single strand or double strand breakage of the DNA molecules. Single strand breakages are amenable to repair. However, double stranded breakages of DNA cannot be repaired and will result in cell death. If the radiation is targeted to cancer cells, they will experience the above reactions and die. This is the basis of the use of radiation for cancer therapy.

As radiation can (almost) equally kill cancer and normal cells, care ought to be taken to minimize damage to normal cells. This is achieved in many ways. For EBRT the definition and placement of irradiation beams is planned so that tumours receive the desired absorbed dose while the absorbed doses for normal tissues are kept to an acceptable level — namely, below the threshold for the appearance of side effects.

An alternative is to use biological targeting molecules to carry the ionizing radiation source to the cancer site. Associating the selective affinity of a biological vector for the tumour with the short range of particles emitted after the radioactive decay of a radionuclide gives the therapy increased selectivity. This mode of therapy is called targeted radionuclide therapy (TRT). Both EBRT and TRT are used for the management of bone metastases [4].

Figure 2 depicts EBRT and TRT. In EBRT, the radiation is delivered from an external source and hence an absorbed dose to other healthy organs and tissues cannot be completely avoided. Thus, radiation induced post-therapy side effects can occur. In TRT, radioactivity is administered systemically, and is expected to accumulate in the target while sparing healthy organs. The absorbed dose for healthy organs and tissues is low because of the specific targeting of the radiopharmaceuticals and the short range of the emitted radiation. Therefore, in principle, TRT ought to induce few side effects.

2.6.2. External beam radiation therapy

In EBRT an intense beam of highly penetrating radiation is focused directly onto the tumour site until the desired absorbed dose is delivered to the tumour. This can be achieved by using gamma radiation emitted by a radioactive source. Currently, an electron beam accelerator (also called LINAC) is used for this purpose. A high intensity ^{60}Co source was previously used for this purpose.

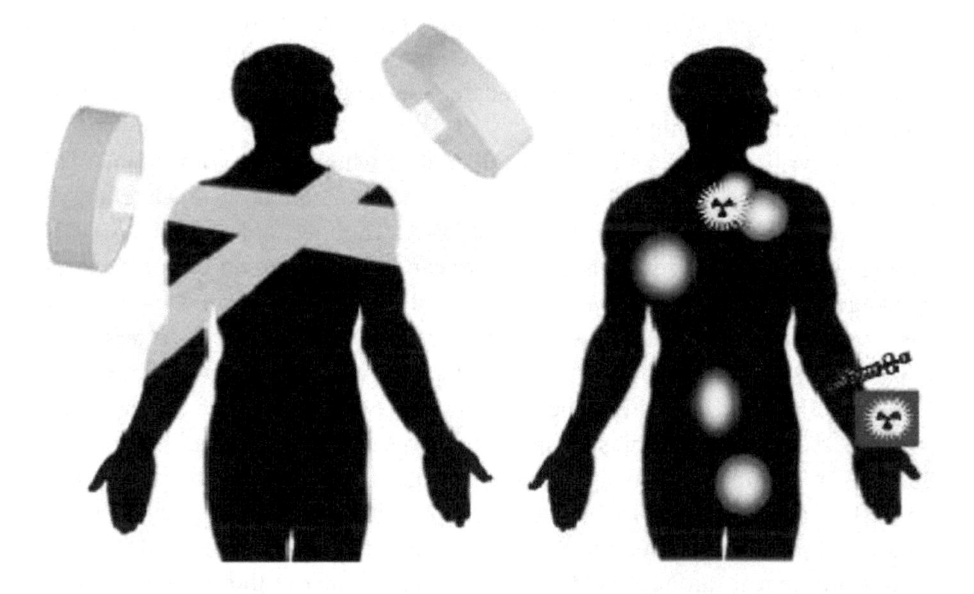

FIG. 2. Use of ionizing radiation for cancer therapy. Left: EBRT in which the radiation is delivered from an external source. Right: TRT in which the radioactivity is administered systemically and accumulates in the target (courtesy of M. Zalutsky, Duke University).

Radiotherapy is widely used for alleviating pain in cancer patients. Radiotherapy is given in either single or multiple dose fractions [37]. An absorbed dose ranging from 30 to 75 Gy is given in the case of prostate cancer patients, depending on the weight of the patient and the tumour load. The main problem with EBRT is the inability to avoid the irradiation of healthy organs, especially when treating widely distributed metastases.

2.6.3. Targeted radionuclide therapy

Metastatic bone pain is also effectively managed by using bone seeking radiopharmaceuticals. Radiopharmaceuticals comprising radionuclides decaying by the emission of short range alpha or beta particles are used in this context. The approved drugs in this category are metastron ($^{89}SrCl_2$), lexidronate (^{153}Sm–EDTMP (ethylene diamine tetramethylene phosphonic acid)) and xofigo ($^{223}RaCl_2$) [6, 38]. This topic will be discussed in detail below.

3. DESIGN OF BONE PAIN PALLIATION AGENTS

3.1. BONE SEEKING RADIOPHARMACEUTICALS

It is important to understand the structure of the bone as well as the related metabolic activities when developing bone targeting radiopharmaceuticals. Bone consists of flexible matrix and bound minerals, which are woven intricately by a group of specialized bone cells. The bone matrix comprises 90–95% collagen fibres, which are elastic, while the remaining 5–10% is made up of ground substances. The matrix is hardened by the binding of inorganic salt calcium phosphate arranged in a chemical form known as calcium HA ($Ca_{10}(PO_4)_6(OH)_2$). This bone mineralization gives the bone strength. It also maintains the delicate balance between calcium and phosphorus, which are essential elements needed for body metabolism.

Bone is continuously formed and destroyed throughout a lifetime. Bone cells called osteoblasts are responsible for bone formation, while osteoclasts are responsible for bone resorption [39]. Both calcium and phosphorus are thus taken up by bones during osteoblastic activity.

Like all other organs and tissues, bone is also susceptible to both primary cancer as well as metastasis of cancers originating elsewhere in the body. During cancer the bone metabolism is altered and the uptake of both calcium and phosphorus increases. Hence, calcium mimics, such as ^{89}Sr and ^{223}Ra as well as radioactive phosphorus, are taken up by bone cancer cells. Both primary and secondary bone cancer cells express specific biochemical markers, some of which can be used as targets to develop bone seeking therapeutic radiopharmaceuticals.

3.2. IN VIVO UPTAKE MECHANISMS OF BONE SEEKING RADIOPHARMACEUTICALS

Different uptake mechanisms are efficiently exploited for the development of bone seeking radiopharmaceuticals. These are discussed briefly in the following sections.

3.2.1. Radionuclides that mimic calcium as bone seeking radiopharmaceuticals

Bone is a great reservoir for body calcium. The uptake of calcium increases significantly during bone metastasis. In 2016, Chakravarty et al. demonstrated the uptake of a long lived β^- emitting radionuclide, ^{45}Ca, in bone [40]. Although the low energy (0.3 MeV) β^- particles emitted by this isotope are suitable for therapy, their long half-life ($T_{1/2}$ = 163 days) is a detrimental factor and hence they are not yet used in therapy. Calcium is a group 2 alkaline earth metal. All elements in this group mimic the properties of calcium and hence are taken up by bone. This is the basis of the use of ^{89}Sr and ^{223}Ra, as both these elements belong to group 2 in the periodic table [41–44]. The radionuclides ^{89}Sr and ^{226}Ra are used in their inorganic salt form as radiopharmaceuticals for bone pain palliation.

3.2.2. Radiopharmaceuticals that are taken up by hydroxy apatite particles

Radioactive phosphorus (^{32}P) in the chemical form of sodium orthophosphate is used as a bone seeking radiopharmaceutical based on its uptake by bone for the formation of calcium HA, which is the major bone mineral. Alterations in the metabolism of phosphate are usually observed in tumours and constant phosphate supply is one of the requirements of metastatic bone cancer cells. A microenvironment rich in inorganic phosphates is highly conducive for the proliferation of osteolytic bone metastasis [45, 46]. Hence, if ^{32}P is administered to patients suffering from bone cancer, it will accumulate more in proliferating cells and the decay energy can be used to kill the cancer cells.

Phosphonate ligands with P–C–P bonds express a strong affinity for osteoblastic bone cells [47]. These bisphosphonates have similar physical and chemical effects to pyrophosphate ligands but are resistant to enzymatic splitting and metabolic breakdown. This is because, unlike P–O–P binding of pyrophosphate, the P–C–P bonding of the bisphosphonates cannot be broken down enzymatically, so their activity is retained in the bone. Using ligands with P–C–P bonds instead of P–O–P bonds represented a genuine breakthrough, which enabled the development of the potent bisphosphonates that are used for therapy of bone disorders [48]. The phosphonate ligand was first utilized in the early 1970s in the preparation of ^{99m}Tc–MDP to image bone [49]. MDP is a diphosphonic acid ligand and it is presumed that two molecules of MDP complex with ^{99m}Tc (Fig. 3). The exact structure of ^{99m}Tc–MDP has not yet been reported. Normal bones can take up ^{99m}Tc–MDP, but uptake increases during bone metastasis.

The use of organic amine phosphonate ligands for the development of bone seeking therapeutic radiopharmaceuticals was reported at the University of Missouri by Troutner and colleagues, who developed ^{153}Sm–EDTMP as a successful therapeutic radiopharmaceutical for bone pain palliation (Fig. 4) [50]. The main advantage of EDTMP is that it has four phosphonate groups and two amine groups, all of which participate in the complexation reaction with metals. Three water molecules occupy the remaining coordination sites of Sm, resulting in a nine-coordinate complex [51]. The patent filed on the above development also proposed that, in addition to ^{153}Sm, several other metallic radionuclides, such as ^{159}Gd, ^{166}Ho, ^{175}Yb and ^{177}Lu, can also be used for the development of bone pain palliation radiopharmaceuticals [52].

A number of other acyclic and cyclic tetraphosphonate ligands were synthesized by Pillai and co-workers at the Bhabha Atomic Research Centre, who studied their complexation with radiometals such as ^{153}Sm, ^{177}Lu, ^{170}Tm, ^{175}Yb and ^{188}Re [53–59]. The phosphonates coordinated to these metallic radionuclides are the bone seekers.

There has also been interest in radiolabelling bisphosphonate ligands conjugated to a bifunctional chelating agent (BFCA) [60, 61]. The metal

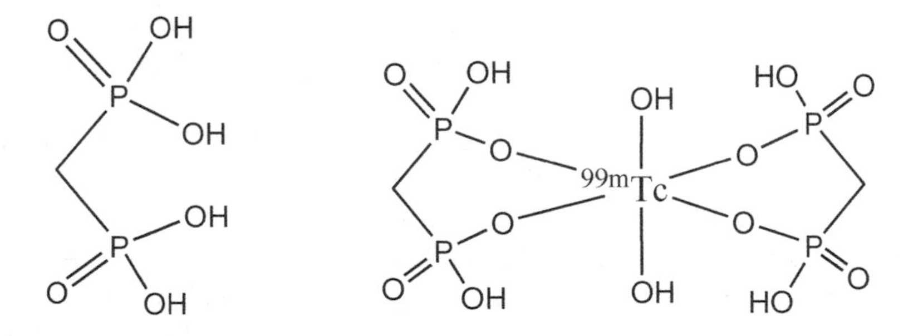

FIG. 3. Proposed structure of ^{99m}Tc–MDP acid (courtesy of M.R.A. Pillai, Molecular Group of Companies).

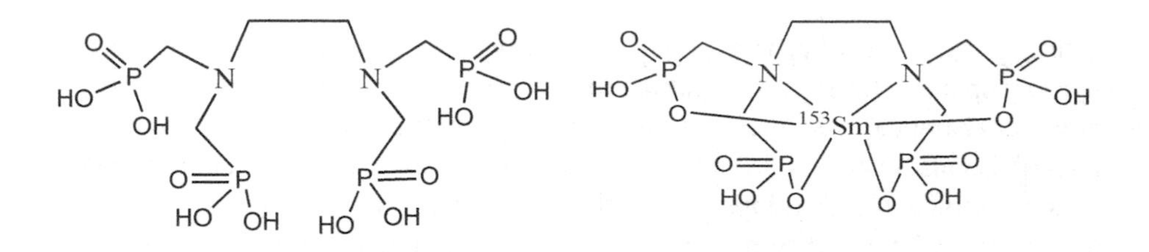

FIG. 4. EDTMP (left) and representative structure of ^{153}Sm–EDTMP (right). A crystallographic study of ^{153}Sm–EDTMP had three water molecules to complete the coordination sphere and five sodium atoms to balance the charge (reproduced from Ref. [51] with permission).

complexation of these modified ligands is through the BFCA, with no involvement of the phosphonate groups for chelation, although crystallographic evidence has not yet been reported. The binding of these BFCA modified tracers with the bone is expected to be better, as the phosphonate groups are not involved in radiometal coordination and are thus free to interact and bind with the calcium HA present in the bone.

3.2.3. Radiopharmaceuticals that are radiolabelled with bifunctional chelating agents (BFCA) conjugated to target seeking molecules

The radiopharmaceuticals described in the previous section use the increased metabolic activity of bone cancer cells. However, a new class of targeting radiopharmaceuticals is being designed based on the overexpression of certain oncoproteins in the cancer cells.

Primary as well metastatic bone cancers overexpress certain specific molecules that can be used as targets for developing bone seeking radiopharmaceuticals. These types of targets include molecules such as receptors and enzymes. Bioactive molecules that bind with the target molecule are used as radionuclide carriers. Biological vectors such as humanized or chimeric antibodies, antibody fragments, peptide receptor ligands, enzyme inhibitors, aptamers, peptidomimetics, non-peptide receptor ligands and DNA analogues can be used for this purpose [2].

Prostate cancer overexpresses an enzyme called PSMA. Radiopharmaceuticals targeting PSMA have been evaluated for a long time as therapeutic products for the management of metastatic prostate cancer. Monoclonal antibodies specific to PSMA labelled with ^{177}Lu were investigated for this purpose but with limited success [62]. The use of inhibitor molecules as carrier vectors to target PSMA has emerged as a very successful strategy for the development of target specific radiopharmaceuticals (TSRs) [63]. PSMA enzyme inhibitor molecules labelled with ^{177}Lu and ^{225}Ac are successful target specific therapeutic radiopharmaceuticals [64, 65].

A BFCA molecule is attached either directly or preferably through a linker molecule so that the biological affinity of the molecule to bind with the target is not compromised. An enzyme inhibitor molecule developed to target the PSMA molecule and used for both diagnosis and therapy of prostate cancer is shown in Fig. 5. The BFCA used in this case is DOTA (1,4,710-tetraazacyclododecane-1,4,7,10-tetraacetic acid), which is capable of binding with several therapeutic radionuclides, including ^{90}Y, ^{177}Lu and ^{225}Ac [64].

3.3. DEVELOPMENT OF BONE SEEKING RADIOPHARMACEUTICALS

There are several things that ought to be considered when developing radiopharmaceuticals, including the selection of a radionuclide and carrier molecule, the optimization of radiolabelling conditions, pre-clinical in vitro and in vivo evaluations, dosimetric estimations and, finally, clinical trials.

3.3.1. Selection of radionuclides

Several radionuclides have the potential to be used for therapy [66, 67]; however, careful selection of the best among them for bone pain palliation is important. The decay mode, half-life, energy and type of emitted particles are important when selecting a radionuclide for bone pain palliation therapy.

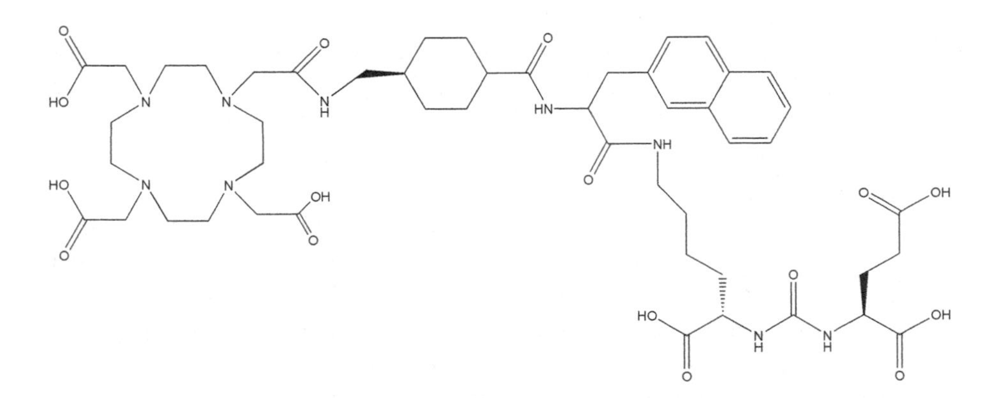

FIG. 5. PSMA enzyme inhibitor molecule modified by adding the BFCA molecule DOTA through a linker moiety (courtesy of T. Das, Bhabha Atomic Research Centre).

3.3.1.1. *Decay mode*

The efficacy of radionuclide therapy depends on the ability to deposit sufficient radiation energy to the cancer cells, leading to their death. Radionuclides that decay by particulate emissions are selected for therapy. Alpha particle, β^- particle and Auger electron emitters are preferred for therapy. These radiations have short path lengths and hence very high LET, and so are able to induce radiation damage to the cancer cells.

3.3.1.2. *Energy of particles*

The energy of the emitted β^- particles is an important consideration for successful metastatic bone pain palliation therapy. Bone marrow, which has the important function of producing both red and white blood cells, is an integral part of the bone. It is necessary to minimize the damage to bone marrow while performing bone pain palliation therapy. A reduction in blood cell counts is one of the major side effects of bone marrow damage due to radiation. Radionuclides emitting low energy β^- particles are preferred for bone pain palliation as they induce less damage to the bone marrow. Blood cell counts are monitored carefully during bone pain palliation therapy to ensure that patients do not suffer from myelotoxicity. The quantity of radiopharmaceutical administered is decided such that the bone marrow dose does not exceed the critical dose limit.

Bone marrow is the most sensitive organ to radiation dose. The radiation dose used for bone palliation ought to be optimal to illicit the requisite onset of palliation action while sparing the bone marrow tissues from irradiation damage. Low to medium energy β^- emitters that permit homogenous distribution of radionuclide in bone are preferred for this purpose.

Guerra Liberal et al. performed studies in which they exposed cells to 11 different radionuclides and estimated the percentage of DNA damage. Their finding was that among the radionuclides studied (^{32}P, ^{89}Sr, ^{90}Y, ^{117m}Sn, ^{153}Sm, ^{166}Ho, ^{170}Tm, ^{177}Lu, ^{186}Re, ^{188}Re, ^{223}Ra), the beta particles from ^{170}Tm and ^{177}Lu induced the highest cell death of all investigated beta particle emitting radionuclides [68].

The energy of the α particles is less critical, as even very high energy α particles will lose their energy completely in 50–100 μm of path length, which is approximately the size of two human cells. The high LET in the case of α particles helps in selective irradiation of the bone cells while at the same time sparing the bone marrow. The LET of α particles is very high, ~80 keV/μm as compared to ~0.2 keV/μm for the β^- particles [69]. Alpha particles can cause irreparable damage to DNA. The main advantage of alpha therapy is that high levels of radiation dose can be delivered to the bone without affecting bone marrow, provided that the radiopharmaceutical concentrates in the bone tissue. Hence, as sufficient radiation dose can be administered without marrow toxicity, α emitters can be used not only for pain palliation, but also for the destruction of metastatic bone cancer.

3.3.1.3. Gamma rays: accompanying radiation

There are advantages and disadvantages to having gamma emission as an accompaniment to particulate emission. As the most penetrative type, γ radiation delivers unnecessary radiation to non-target organs and tissues and at the same time does not add any advantage for localized therapy. The absence of γ radiation allows therapy to be performed as an outpatient procedure without the need for hospitalization. Further, bystanders and the public are not exposed to radiation when pure β^- or α emitters are used for therapy.

However, the emission of gamma photons in the diagnostically useful energy range of 80–200 keV is desirable. In such cases, the pharmacokinetics as well as in vivo localization of the radiopharmaceutical as a function of time can be studied using SPECT imaging. The information collected from these studies is useful for carrying out dosimetry estimations. The emission of photons with low abundances, preferably $<10\%$, is preferred in the case of β^- emitting radionuclides to ensure a minimal increase of the radiation dose burden for the patient. A low abundance of gamma emission also helps in early discharge of patients after therapy. In the case of α particle emitting radionuclides, the amount of radionuclide used in therapy is very low and of the order of ~3.7–7.4 MBq (100–200 μCi). Hence, it is essential to have a higher abundance of gamma rays such that imaging can be performed.

3.3.1.4. Half-life

Another very important characteristic for the selection of a radionuclide for targeted therapy is the physical half-life of the radionuclide. Long lived radionuclides are preferable, as they deliver their radiation dose over an extended period of time and hence the cumulative radiation dose delivered is higher. For example, ^{153}Sm ($T_{1/2}$ 47 h) will deliver a much lower dose compared to ^{89}Sr ($T_{1/2}$ 50.5 days) while using the same amount of radioactivity. Hence, radiopharmaceuticals made with short lived radionuclides need to be administered in higher amounts. Very long lived radionuclides will have the problem that the onset of pain relief may begin only after a relatively long time. For bone pain palliation, radionuclides with half-lives ranging from a few hours (^{188}Re; $T_{1/2}$ = 16.9 h) to several months (^{170}Tm; $T_{1/2}$ = 120 d) have been proposed and used.

An ideal radionuclide to be used widely for bone pain palliation ought to have a half-life that allows global distribution of the products from the production sites. Radioisotope production through radiochemical processing of an irradiated target, radiopharmaceutical preparation, formulation, quality control, dispatch and then delivery to a nuclear medicine department are required before the radiopharmaceutical is made available at the patient's location for treatment. A radionuclide with a longer half-life is advantageous for this, as in many places isotope production and radiopharmaceutical production facilities are located in different places. A long half-life will also reduce decay loss, making it economically more viable. This is one of the reasons why many nuclear medicine physicians prefer $^{89}SrCl_2$ over ^{153}Sm–EDTMP for clinical use, although the clinical efficacy of both products is well established.

3.3.1.5. Logistics of production

The logistics of production is an important consideration when selecting a radionuclide. Even when the nuclear characteristics are highly favourable, the radionuclide will not be useful unless it can be prepared in adequate quantities with the desired quality. Hence, radionuclides that are prepared through nuclear reactions with high neutron absorption cross-sections (σ) are preferred. Among the radionuclides suitable for bone pain palliation, ^{32}P, ^{153}Sm, ^{166}Ho, ^{170}Tm and ^{177}Lu can be prepared relatively easily, whereas ^{89}Sr and ^{117m}Sn are more difficult to produce [70].

3.3.1.6. Specific activity

The specific activity achievable for the radionuclide is a very important consideration when selecting a radionuclide for therapy. No carrier added (NCA) radionuclides have advantages for use in radionuclide therapy as the specific activity of the radiolabelled carrier molecules will be high and will remain unchanged during storage. Medium specific activity radionuclides will suffice for the preparation of the bone pain palliation radiopharmaceuticals discussed in Sections 3.2.1 and 3.2.2. High specific activity radionuclides are mandatory in targeted therapy using carrier molecules such as the peptides and inhibitor ligands discussed in Section 3.2.3.

3.3.2. Selection criteria for carrier molecules

The selection of the most appropriate carrier molecule for the development of a bone seeking radiopharmaceutical is also a difficult task. Elaborate experimentation will be essential before the best carrier molecule is identified. The properties to consider when choosing carrier molecules include the following:

(a) The carrier moiety ought to have high affinity and specificity for the target. The affinity of the carrier molecule ought to remain unchanged after complexation with the radiometal.
(b) The carrier molecule ought to show preferential accumulation in cancer cells rather than in normal cells. It ought to exhibit prolonged retention in the lesions and at the same time rapid washout from non-target organs/tissue.
(c) It ought to be possible to label the carrier molecule or the chosen ligand with the selected radionuclide using simple radiochemical reactions efficiently, resulting in a stable product.
(d) The radiolabelling protocol ought to be simple and convenient, so that the process can be easily adopted in a hospital radiopharmacy.
(e) The carrier moiety by itself or after modification with the BFCA ought to be able to form a complex with the radiometal at a high yield so that post-preparation purification is avoided.
(f) The formed radiometal complex ought to be thermodynamically stable and kinetically inert. The radiometal complex ought to have high stability under physiological conditions.
(g) The carrier moiety ought to be non-toxic and non-immunogenic.
(h) The carrier moiety ought to have high radiation stability, so that it does not undergo degradation when radiolabelled with the high amount of radioactivity needed for therapy.
(i) The synthesis protocol of the carrier molecule ought to be easy to scale up; alternatively, it ought to be available at a reasonable price.

The most commonly used carrier moieties for developing bone pain palliation agents carry two or more phosphonic acid groups in their structure [71]. Several types of phosphonic acids have been used as carrier vectors for a variety of radionuclides in the quest to develop potential bone pain palliation agents [50–59, 72]. The selection of proper organic moieties bearing the phosphonic acid group is important as it can significantly alter the delivery, retention, binding and clearance of the radiopharmaceuticals. In this regard, the use of polyamino polyphosphonic acid derivatives in developing agents for bone pain palliation requires special mention. Although the exact mechanism of bone uptake of phosphonates from blood is not clearly understood, it is generally believed that during the endocytic processes of osteoclasts, phosphonates are internalized, as these are available on the mineral surfaces of the bone matrix when active remodelling processes occur [73].

3.3.3. Radiolabelling

As discussed in the previous sections, the radiopharmaceuticals used for bone palliation fall into three different categories and the radiolabelling concerns are different for each of them, as discussed below.

3.3.3.1. Radiopharmaceuticals used as inorganic salts

There are three bone seeking radiopharmaceuticals in this category, using the radionuclides ^{89}Sr, ^{223}Ra and ^{32}P. There are no radiolabelling issues while using these types of radiopharmaceuticals, as they are used in their inorganic salt forms as strontium chloride ($^{82}SrCl_2$), radium chloride ($^{223}RaCl_2$) and sodium orthophosphate ($Na_3{}^{32}PO_4$). The only consideration for these types of radiopharmaceuticals is to maintain the desired physiological conditions such that they are suitable for administration to patients. There are no major stability concerns as each product is in its most stable form.

3.3.3.2. Radiopharmaceuticals based on amine phosphonate ligands

The nitrogen and oxygen atoms present in amine phosphonate ligands are electron donors and hence capable of complexing with most of the radiometals used for bone pain palliation. However, the radiolabelling conditions have to be optimized carefully to form complexes with high yields and the greatest stability. The ligand concentration and the pH, time and temperature of the reaction are important parameters to obtain the highest complexation yields and they must all be optimized. The ligand requirements for the formation of stable complexes will vary depending upon the ligand and the specific activity of the radiometal used. Cyclic ligands such as DOTMP (1,4,7,10-tetraazacyclododecane-1,4,7,10-tetramethylene phosphonic acid) (Fig. 6) with four amine and four phosphonate groups require less ligand to make radiometal complexes as compared to acyclic diamine tetraphophonic acid ligands. However, the ring size of the ligand is important to provide stable complexes with different radiometals. For example, ^{186}Re forms stable complexes with the 14-member macrocyclic CTMP ligand but the lanthanides do not. On the other hand, the 12-member macrocyclic DOTMP ligand gives high complexation yields with lower ligand concentrations with most of the lanthanide radiometals studied [55, 56].

3.3.3.3. Radiopharmaceuticals with BFCA conjugated carrier molecules

The use of a BFCA chelating agent to radiolabel biomolecules was a novel concept introduced in the 1980s by Hnatowich et. al., who used a DTPA (diethylenetriaminepentaacetic acid) anhydride to add a chelating agent to albumin [74]. A large number of BFCAs were synthesized after that to radiolabel specific radionuclides [75]. Figure 7 shows some of the common BFCA molecules used to modify the carrier molecules.

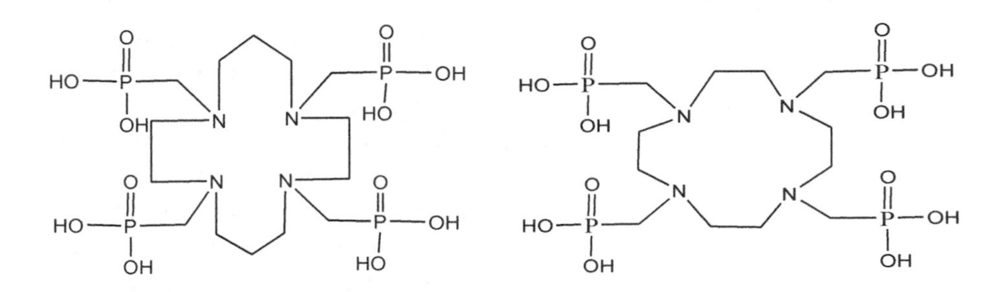

FIG. 6. Tetramine tetraphosphonic acid ligands used for making bone seeking radiopharmaceuticals. CTMP (1,4,8,11-tetraazacyclotetradecane-1,4,8,11-tetramethylene phosphonic acid) (left) and DOTMP (right)(courtesy of T. Das, Bhabha Atomic Research Centre).

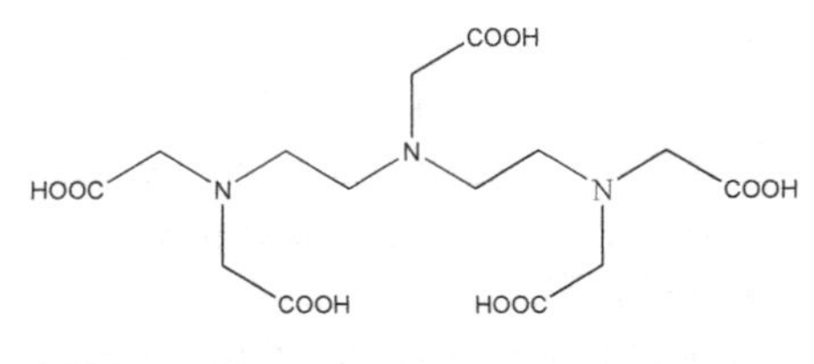

DTPA

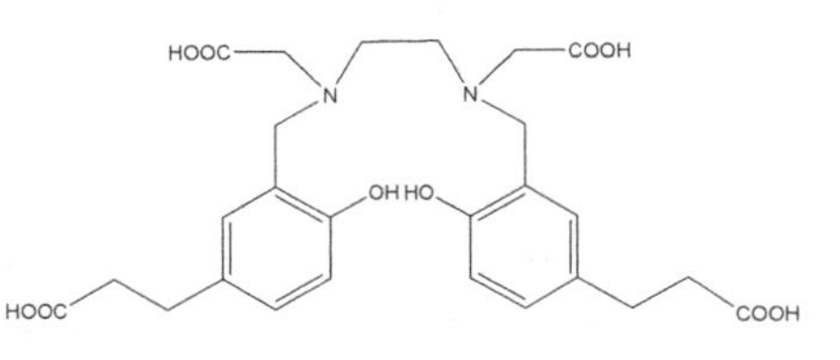

HBED-CC

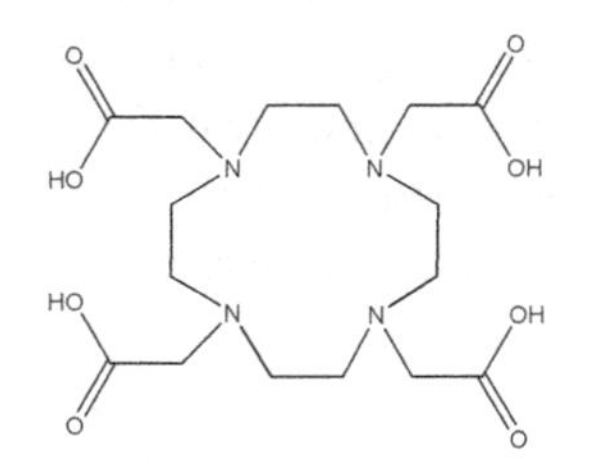

DOTA

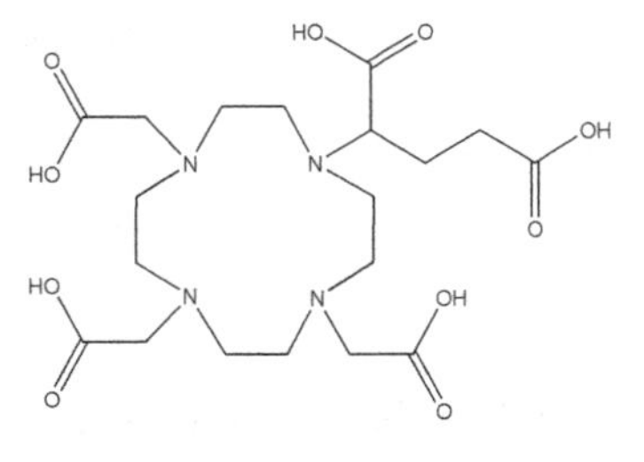

DOTAGA

FIG. 7. Bifunctional chelating agents used for radiolabelling metallic radionuclides (courtesy of T. Das, Bhabha Atomic Research Centre).

3.3.4. Stability of radiometal complexes

The stability of the radiometal complexes formed is critical in radiopharmaceutical chemistry for human administration; the complexes are only used at micromolar to nanomolar levels. Thermodynamic stability is generally considered to be the most important factor in inorganic chemistry. Blood is an extremely complex medium with a variety of binding proteins, chelating moieties and elements, hence under in vivo conditions, when radiometal complexes are present at high dilution, kinetic stability also becomes important. The thermodynamic stability constants are the ratio of the kinetic on rate divided by the kinetic off rate. It is possible to have a high thermodynamic stability constant and a high kinetic off rate (e.g. EDTA complexes of the lanthanides), resulting in unstable complexes. Increasing the denticity and/or adding a macrocycle increases the kinetic stability of the resulting complexes.

Lanthanides such as samarium and lutetium are involved in forming ionic bonds to their ligands. Lanthanides are considered to be hard Lewis acids with high charge density and thus prefer to bind to hard anionic Lewis bases, which are ligands containing hard donors such as oxygen. The most common coordination number for the lanthanides is 9, although 12 coordinate complexes for some of the early lanthanides are known. Multidentate chelators, preferably octadentate, are preferred for preparing kinetically stable lanthanide complexes. Additional stability can be achieved by incorporating macrocyclic chelators. For example, both DTPA and DOTA are octadentate chelators and form complexes with the various lanthanides. However, the macrocyclic DOTA chelator generates kinetically more stable complexes than the acyclic DTPA chelator [76].

Radiopharmaceuticals based on $^{186/188}$Re pose interesting challenges. Rhenium has multiple oxidation states available, and in fact complexes in oxidation states −1 through +7 have been reported [77]. Each oxidation state has its own chemistry and redox stability, and this ought to be considered in addition to thermodynamic and kinetic stability. Rhenium complexes are more readily oxidized to perrhenate than the analogous technetium complexes. The synthesis of all $^{186/188}$Re complexes ought to begin with perrhenate (the chemical form available following production and separation), which has the Re in its highest oxidation state (+7). In order to complex the Re with a ligand, it is necessary for the Re to be reduced to a lower oxidation state (generally +5, +3 or +1).

3.3.5. Radiolabelling conditions

Radiolabelling methods are specific for each radiometal. There could be some similarity, as observed with the different radiolanthanides. The complexation reactions are highly sensitive to pH and depend on the radiometal as well as the chelate used. Lanthanide radionuclides are supplied in acidic solution, as these metals will undergo hydrolysis to insoluble hydroxide or oxide forms at higher pH. Lanthanide complexation with phosphonate ligands is usually performed in bicarbonate buffer at pH~9 [55–59, 78, 79].

Complexation of radiolanthanides with polyaminopolycarboxylate ligands such as DOTA is performed at pH4–5 [80]. Different buffer solutions can be used to maintain the appropriate pH for the reaction. Usually, acetate or ascorbate buffers are used to make radiolanthanide complexes with the DOTA ligand. The ligand concentration is varied to obtain the best complexation yields. The quantity of ligand required will depend on the specific activity of the radiometal used and will be in molar excess to the radiometal. A minimum metal : ligand (M : L) ratio is essential to obtain stable complexes. It is also known that cyclic ligands form stable complexes with a much lower M : L ratio than acyclic ligands. Reaction time and temperature are the other parameters to be optimized.

Rhenium-186/188 ought to be reduced from perrhenate, the chemical form available following production/separation, in order to complex with either a phosphonate or a BFCA with an appended phosphonate or other targeting moiety. Stannous ion is frequently used as the reducing agent and the radiolabelling is carried out under acidic conditions, since Re is highly susceptible to oxidation back to perrhenate at higher pH. The pH is adjusted to near neutral following radiolabelling to make it suitable for injection.

The stability of the radiometal complexes formed is very important. The radiometal complex is incubated under different environments, such as in saline or human serum, and the radiochemical purity is determined at different time intervals. If the stability is poor, further optimization experiments are carried out to obtain complexes with higher stability. The radiolabelling protocol ought to be optimized thoroughly to produce an agent with the highest radiochemical purity. A robust protocol is developed to assure high repeatability in complexation yields.

Quality control studies involve estimation of radionuclidic purity, radiochemical purity and chemical purity. Radionuclidic purity is often estimated by gamma ray spectrometry whenever there is gamma emission. If the radionuclide is sourced from a commercial supplier, the radionuclidic purity will be ensured and details will be provided. Radiochemical purity is determined by chromatographic techniques. While high performance liquid chromatography (HPLC) will be used for estimating radiochemical purity at developmental stages, it is not practical to use HPLC on a regular basis at a hospital radiopharmacy. Paper chromatography or silica gel thin layer chromatography, which can be completed in a few minutes, are used in a nuclear medicine department. Chemical purity is generally ensured, depending on the raw material used, and is estimated by a test developed according to the chemical impurity. In order to use the product for clinical studies endotoxin and sterility tests ought to be carried out as well. Finally, the product ought to have the right physiological conditions for parenteral administration, which is generally performed in saline solution at near neutral pH.

3.4. PRE-CLINICAL STUDIES

The radiolabelled agent used for bone pain palliation ought to preferentially accumulate in the skeleton with a high lesion to normal bone accumulation ratio. The accumulated radioactivity ought to be retained in the bone without any further redistribution for as long as possible. The non-accumulated activity ought to exhibit fast clearance from the body in order to reduce the radiation dose burden to the non-target organs. Therefore, any new agent ought to undergo rigorous in vitro evaluation and its potential ought to be thoroughly evaluated in vivo in appropriate animal models. It is desirable to evaluate the accumulation and retention of the radiolabelled agent in normal as well as disease induced animals prior to initiating the clinical studies. Toxicological studies ought to be conducted, if information is not explicitly documented for the carrier moiety and the radionuclide used for developing the new product. All experiments involving animals will need clearance from local animal ethics committees.

3.5. CLINICAL TRIALS

If an agent is found to be suitable after the completion of all the necessary pre-clinical and toxicological studies, preliminary clinical studies can be initiated in a limited number of patients. The study ought to be continued in a larger population of patients to determine the effective dose and efficacy of the agent in controlling the metastatic bone pain subject to the realization of favourable results in the preliminary clinical studies. Clearance from the respective local institutional ethical committee is mandatory for conducting all studies involving patients. Clinical trials are very involved and expensive. As the market for radiopharmaceuticals is relatively small, many commercial companies have limited interest in conducting large scale clinical studies. Hence, several new tracers that have been developed have not moved from in vitro animal studies to the clinic.

The IAEA has taken the initiative and helped several Member States through a CRP to coordinate a clinical trial of one of the bone pain palliation radiopharmaceuticals, ^{177}Lu–EDTMP. The studies were performed in multiple stages. Phase I involved the use of a tracer dose (185 MBq/5 mCi) to determine the pharmacokinetics and bone uptake of the tracer in a small number of patients, followed by a dose escalation study, in which the maximum tolerated dose was estimated [7]. Following this, a phase II clinical trial involving a therapeutic dose was performed in a few Member States [8, 9]. Based on these results, approval for the manufacture and supply of ^{177}Lu–EDTMP for human use has been granted in certain Member States.

3.6. APPROVAL

Every radiopharmaceutical needs to be approved by the drug licensing agency of respective countries before it can be used as a regular product for clinical use. The drug licensing agencies in each country will have their own rules and guidelines, but the data obtained from the clinical trials, if they unequivocally demonstrate the efficacy and safety of the new radiopharmaceutical, are the principal basis for granting such approval.

3.7. GOOD MANUFACTURING PRACTICES FOR RADIOPHARMACEUTICAL PRODUCTION

Radiopharmaceuticals for bone pain palliation are prepared either as finished products or using freeze-dried kit formulations. In the latter case, the radionuclide is added to the kit at the nuclear medicine department to formulate the clinical dose. The preparation of the finished products as well as kits ought to be in accordance with the standards established for radiopharmaceutical manufacturing. The standard guidelines for the manufacture of radiopharmaceuticals are described below, in brief.

According to the World Health Organization, radiopharmaceuticals can be classified as a radiopharmaceutical preparation, a radionuclide generator, a radiopharmaceutical precursor and a kit for radiopharmaceutical preparation [81]. The term 'small scale' radiopharmaceutical has been introduced for any in-house PET, SPECT or therapeutic radiopharmaceutical prepared on a small scale [82].

Good manufacturing practices constitute a well known system to ensure the quality of the pharmaceutical product for the benefit of the patient. Radiopharmaceuticals for human application ought to comply with the specifications of a regulatory framework, although with certain peculiarities. The manufacture and control processes are operations that involve potential risks inherent to their radiation characteristics and the administration pathway. It is well known that radiopharmaceuticals do not have a pharmacological effect, but most of them are administered intravenously. Accordingly, they ought to meet the requirements for sterile pharmaceuticals.

The good manufacturing practices for manufacturers describe the requirements that the processes ought to meet, from the design of the installation through the development, preparation, production, handling, packaging, conditioning, stability, quality control, storage and distribution of radiopharmaceuticals in each country. Good manufacturing practices aim to ensure that these factors consistently meet the quality requirements of the product for patient administration [83, 84].

The first step in implementing good manufacturing practices consists of identifying the process that needs to be kept under control. All of these processes ought to be well documented in standard operating procedures (SOPs). The SOPs ought to be integrated into a documentation system that ensures their correct management. The written, reviewed and authorized procedures ought to specify all the aspects (step by step) involved in the manufacturing or preparation of the radiopharmaceutical.

All suppliers of radioactive raw materials ought to be evaluated and approved. The received product ought to be verified according to documented specifications and be provided with a certificate of compliance. Its storage (if applicable) and handling also ought to be documented. Specific records ought to be kept allowing traceability of the received quantities against the supplied quantities. Handling of radioactive materials ought to be performed by authorized personnel. All of the critical operations ought to be verified and carried out according to authorized SOPs.

All instruments used for measurement ought to be calibrated and the areas and systems ought to be qualified. Appropriate preventive maintenance, calibration and qualification plans and programmes ought to be carried out for the facilities and for the equipment used in the manufacturing or preparation of the radiopharmaceutical.

Automated radiochemical devices ought to be located in shielded hot cells. Prior to the preparation of an in-house radiopharmaceutical, the operator ought to conduct pre-established checks. Hot cells ought to be tested for pressure to ensure the correct performance of the ventilation system. Radiation monitoring devices also ought to be verified. The dispensing systems to prepare the radiopharmaceutical doses ought to be cleaned and the sterile disposables (tubing, filters, syringes, needles, valves and other materials) ought to be replaced. The high efficiency particulate air filter integrity test is mandatory within the hot cell for dispensing [82].

Each batch of manufactured product ought to be controlled from the beginning of the process to guarantee that the materials are fully controlled. An SOP ought to establish the parameters and control activities for the process. Process controls ought to include all measurable parameters involved in the manufacture or preparation (temperature, radioactivity levels, pressure, time).

The results of the tests and analyses carried out during the process ought to be recorded or annexed to the production batch. The intermediate and final yields ought to be registered and compared against their limits. If any deviation from the written instructions occurs, it ought to be documented, and an investigation ought to be carried out.

The aseptic areas ought to be classified and ought to be periodically controlled to meet the correct classification. Environmental monitoring ought to be carried out according to an authorized programme. Immediately after the end of the process, the product ought to be conditioned and kept in controlled conditions until its delivery.

The quality control equipment ought to be qualified periodically using appropriate standards. Analytical weight balances, multichannel analysers, dose calibrators and radio-chromatogram scanners ought to be calibrated and maintained according to authorized programmes.

Validated testing methods (when applicable) and test methods ought to be available. Quality control ought to determine the release of the finished product in accordance with procedures. The conditions of transport ought to be written in order to preserve the characteristics of the radiopharmaceutical. Complaint handling, internal audits, external audits and SOPs for recall also ought to be available.

In addition to the drug regulations, the radiopharmaceutical manufacturing ought to meet the radiological safety regulations established by national, regional or international regulators. As emerging radiopharmaceuticals for therapy or diagnosis have been reported, and some will eventually be translated to clinical practice, it will be imperative to have a clear regulatory framework for the manufacturer and the authorities in order to optimize the authorization times for marketing.

Several pharmacopoeias (USP, European, International, Brazilian, British, Indian, Mexican, etc.) include monographs on radiopharmaceuticals, which contain a description and the tests to perform on them in order to allow the release of the final product. A few countries have also framed national rules on this issue. Countries that do not have national rules can apply international guidelines.

4. RADIONUCLIDES FOR BONE PAIN PALLIATION

4.1. GENERAL PRINCIPLES FOR THE PRODUCTION OF RADIONUCLIDES

Nuclear reactors or charged particle accelerators such as cyclotrons are used for the production of radionuclides. Radionuclides produced in cyclotrons are neutron deficient and hence mostly decay by positron emission or electron capture. Positron emitters have no role in targeted radionuclide therapy. A few of the radionuclides decaying by electron capture could also emit Auger electrons, which are therefore suitable for therapy. There are a few radionuclides, such as ^{225}Ac, that can be produced in a cyclotron and find applications in radionuclide therapy [85].

Most of the radionuclides used for bone pain palliation therapy are produced in nuclear reactors [86]. These are neutron rich radionuclides decaying by emission of β^- particles. Another means of obtaining radionuclides for radiopharmaceutical preparations is the separation of a decay radionuclide from a long lived parent radionuclide [87]. This can be done either at a manufacturing site or by providing the parent radionuclide as a ready to use radionuclide generator [88]. For example, ^{90}Y is isolated from ^{90}Sr ($T_{1/2}$ 30 years) at the manufacturing site and shipped as a radionuclide for radiopharmaceutical preparation or supplied as a finished product [89]. On the other hand, ^{188}Re used for TRT is obtained from a ready to use ^{188}W/^{188}Re generator [90].

Some of the alpha emitting radionuclides used for therapy are isolated from their very long lived alpha emitting parent radionuclides in the uranium, thorium or neptunium series. Some of these alpha emitting radionuclides may be further irradiated in a nuclear reactor or cyclotron to produce radionuclides useful for bone pain palliation [85].

4.1.1. Cyclotron production of radionuclides

In the case of cyclotron production of radionuclides, suitable targets are irradiated with high energy particles such as protons, deuterons and alpha particles. Depending on the energy of the induced particles, usually one or more neutrons are emitted from the nucleus of the target atom. The typical reactions where neutrons are emitted are denoted as (p,*x*n), (d,*x*n) or (α,*x*n) where *x* is the number of emitted neutrons. It is also possible to emit particles such as protons and alpha particles along with neutrons when irradiation is performed with high energy particles in certain targets. The produced radionuclide needs to be separated from the target using suitable chemical separation procedures. There are IAEA publications providing detailed descriptions of the production of radionuclides using cyclotrons [91, 92].

4.1.2. Reactor production of radionuclides

The major source of radionuclides for bone pain palliation is the nuclear reactor. In a nuclear reactor, suitable target materials are mostly irradiated at thermal neutron fluxes ranging from 10^{13}–10^{15} $n \cdot cm^{-2} \cdot s^{-1}$. The most common nuclear reaction is (n,γ) where a neutron enters the target atom resulting in the formation of an excited atom. The excited atom comes to the ground state by emission of a gamma photon.

The activity of the product formed in such a nuclear reaction is given by the equation:

$$A = Nsf\left(1 - e^{-lt}\right)\text{Bq} \quad (1)$$

where

N is the number of target atoms;
σ is the cross-section (1 barn = 10^{-28} m^2 or 10^{-24} cm^2);
ϕ is flux in the irradiation position;
λ is the decay constant, which is 0.693/T½;

and t is the time of irradiation.

Generally, only the cross-section (σ) for thermal neutrons is taken for yield calculations. However, depending on the irradiation position in the reactor, there will be a component of neutrons with energy in the epithermal (0.25 eV–5.0 keV) region. Some of the nuclear reactions have significant cross-sections in the epithermal region and hence the epithermal cross-section as well as resonance integrals need to be factored while performing yield calculations.

The radionuclides prepared by direct (n,γ) nuclear reaction will often have low specific activity as there will be a significant amount of target nuclide present as a carrier in the final preparation. The specific activity of the product formed will also depend on the neutron activation cross-section (σ), the neutron flux (ϕ), which position the irradiation is carried out at and the time of irradiation (t).

A typical example is the production of ^{177}Lu by irradiating either natural Lu or an enriched ^{176}Lu target by the ^{176}Lu(n,γ)^{177}Lu reaction. The radiochemical processing of such targets is simple as there are no other elemental impurities present. Once irradiation is complete, the target is recovered and processed in a hot cell. Usually, simple steps of dissolution, removal of acid, adjustment of pH, and so on will recover the activity [93].

Another method of production is when neutron irradiation results in the formation of a short lived radionuclide. This radionuclide then decays to produce another longer lived radionuclide that is useful for therapy. An example for this type of process is the production of ^{177}Lu by irradiation of an enriched Yb (in ^{176}Yb) target, which produces the short lived ^{177}Yb ($T_{1/2}$ 1.91 h) and then decays to ^{177}Lu, as seen in Eq. (2) [94].

$$^{176}\mathrm{Yb}(\mathrm{n},\gamma)^{177}\mathrm{Yb}\xrightarrow{\beta-}{}^{177}\mathrm{Lu} \quad (2)$$

Following irradiation, the target needs to be stored for the decay of the first radionuclide and then chemically processed to separate the bulk target material from the useful radionuclide. The radionuclides produced through this route are called NCA and will have specific activity close to their theoretical specific activity.

An alternative to thermal neutron reactions are fast neutron reactions, which use high energy neutrons, resulting in reactions where particles are emitted. A typical example is the nuclear reaction $^{59}Co(n,p)^{59}Fe$. The cross-sections for fast neutron reactions are generally much lower than for thermal neutron reactions. Phosphorus-32, used for bone pain palliation, is prepared by the nuclear reaction $^{32}S(n,p)^{32}P$ [95]. An elemental sulphur target is irradiated in a nuclear reactor at positions with high fast neutron fluxes. A chemical purification will be needed to recover the radionuclide from the bulk target.

Strontium-89 can also be prepared by irradiating yttrium with fast neutrons. Yttrium is mononuclidic and hence a natural yttrium target can be used for production by using the following nuclear reaction: $^{90}Y(n,p)^{90}Sr$ [96].

4.1.3. Generator production of radionuclides

Radionuclides can also be obtained from a radionuclide generator wherein a long lived radionuclide decays to a short lived decay product [88]. A typical example of such a generator is that used for the preparation of ^{188}Re [90]. A major advantage of the radionuclides prepared from generators is that they are NCA and have the highest specific activity. The parent radionuclide used for making the generator needs to be produced in a cyclotron or a nuclear reactor. Table 1 lists some of the therapeutic radionuclides that are produced using radionuclide generator systems.

TABLE 1. GENERATOR PRODUCED RADIONUCLIDES USEFUL FOR THERAPY

Radionuclide generator	Parent radionuclide	Half-life of the parent	Decay radionuclide	Half-life of the decay
Actinium-225/ bismuth-213	Actinium-225	10.0 d	Bismuth-213	45.59 min
Calcium-47/ scandium-47	Calcium-47	4.53 d	Scandium-47	3.345 d
Lead-212/bismuth-212	Lead-212	10.64 h	Bismuth-212	60.55 min
Strontium-90/yttrium-90	Strontium-90	28.79 years	Yttrium-90	64.10 h
Tungsten-188/ rhenium-188	Tungsten-188	69.4 d	Rhenium-188	16.98 h

4.1.4. Isolation from long lived actinides

Some useful therapeutic radionuclides, especially those decaying by alpha particle emission, can be isolated from long lived natural or artificially produced actinide elements. There is a large inventory of actinides, which are the by-products of the operation of nuclear reactors and fuel processing plants. One radionuclide prepared through this route is ^{225}Ac, which is used for alpha therapy. Thorium-229 ($T_{1/2}$ 7880 years) is a decay product of ^{233}U ($T_{1/2}$ 160 000 years). Uranium-233 and ^{229}Th are in secular equilibrium. Thorium-229 is separated from the ^{233}U inventory and is used for the separation of ^{225}Ac. Thorium-229 decays to ^{225}Ra ($T_{1/2}$ 14.9 days). Radium-225 and ^{225}Ac are in transient equilibrium and hence the latter can be separated at regular intervals for clinical use.

4.2. PRODUCTION OF β^- PARTICLE EMITTING RADIONUCLIDES

Table 2 presents the β^- particle emitting radionuclides that have been used or are potentially useful for the preparation of radiopharmaceuticals for bone pain palliation.

TABLE 2. LIST OF β^- EMITTING RADIONUCLIDES USEFUL FOR BONE PAIN PALLIATION

No.	Radionuclide	Half-life	Maximum energy of β^- particles in KeV (abundance)	Energy of gamma emissions in KeV (abundance)
1.	Phosphorus-32	14.26 d	1711 (100%)	None
2.	Strontium-89	50.53 d	1497 (100%)	None
3.	Samarium-153	46.27 h	808 (17.5%) 705 (49.6%) 635 (32.2%)	103 (28.3%), 70 (5.25%)
4.	Lutetium-177	6.734 d	498 (78.6%) 385 (9.1%) 176 (12.2%)	208 (11.0%), 113 (6.4%)
5.	Rhenium-186	90.64 h	1069 (80.0 %) 932 (21.54%) 581 (5.78%)	137 (8.6%)
6.	Rhenium-188	16.98 h	2120 (71.1%) 1965 (25.6%)	155 (14.9%)
7.	Tin-117m	14 d	130 (64.9%) 150 (26.2%)	0.159 (86.4%)
8.	Thulium-170	128.6 d	968 (81.6%) 883 (18.3%)	84 (3.26%)
9.	Ytterbium-175	4.185 d	470 (86.5%) 356 (3.3%) 73 (10.2%)	396 (6.5%) 282 (3.10%)
10.	Holmium-166	26.83 h	1854 (50.0%) 1774 (48.7%)	81 (6.2%)
11.	Yttrium-90	64 h	2272 (99.983%)	Nil
12.	Terbium-161	6.88 d	593 (10.0%) 567 (10.0%) 518 (66.0%) 461 (26.0%)	74 (9.8%), 49 (14.8%),

TABLE 2. LIST OF β^- EMITTING RADIONUCLIDES USEFUL FOR BONE PAIN PALLIATION (cont.)

No.	Radionuclide	Half-life	Maximum energy of β^- particles in KeV (abundance)	Energy of gamma emissions in KeV (abundance)
13.	Erbium-169	9.4 d	351 (55.0%) 342 (45.0%)	Insignificant

4.2.1. ^{32}P

Phosphorus-32 decays with a half-life of 14.26 days and emits β^- particles with an $E_{\beta(max)}$ of 1711 keV. It is a pure β^- emitter with no γ emission. Phosphorus-32 is one of the earliest artificially produced radionuclides, having its origin in 1936 at the Berkeley cyclotron where E.O. Lawrence produced ^{32}P by $^{31}P(d,p)^{32}P$ nuclear reaction. However, meaningful quantities for large scale therapy cannot be produced through this route. Nevertheless, the first clinical study with ^{32}P leading to the use of artificially produced radionuclides for medical application was performed with ^{32}P prepared through this route. This route of production has now been abandoned.

Phosphorus-32 is prepared in nuclear reactors using the following nuclear reactions:

$^{31}P(n, \gamma)^{32}P$, 100% abundance, σ 0.172 b
$^{32}S(n,p)^{32}P$, 100% abundance, σ 0.068 b

Vimalnath et al. reported the preparation of ^{32}P using the above two routes [97]. The production processes are described in brief below.

4.2.1.1. Production of ^{32}P using red phosphorus

Red phosphorus is irradiated at a neutron flux of ~7.5×10^{13} $n \cdot cm^{-2} \cdot s^{-1}$ for a period of 6 days. Radiochemical processing is performed by dissolving the irradiated red phosphorus in concentrated nitric acid. After complete dissolution, the residue is dissolved in HCl and heated to near dryness. The residue is reconstituted in ultrapure water and heated to near dryness; this step is repeated three times to remove the acids completely. The contents are finally dissolved in ultrapure water to yield ^{32}P-orthophosphoric acid ($H_3{}^{32}PO_4$). The specific activity of the radionuclide produced is adequate for the preparation of bone seeking radiopharmaceuticals.

4.2.1.2. Production of ^{32}P using sulphur targets

An elemental sulphur target is irradiated for 60 days at a fast neutron flux of ~8.9×10^{11} $n \cdot cm^{-2} \cdot s^{-1}$. Radiochemical processing involves the initial distillation and removal of sulphur, leaving behind the ^{32}P. After ensuring complete distillation of the sulphur, the remaining contents are dissolved in 0.05 N supra-pure HCl and further distilled at ~60–70°C, collecting $H_3{}^{32}PO_4$ in the receiver flask. $H_3{}^{32}PO_4$ thus prepared is purified by passage through a cation exchange resin to remove aluminium and other cationic impurities. The eluate containing $H_3{}^{32}PO_4$ is collected and heated to dryness to remove the excess HCl. The residue is reconstituted in 0.05 M supra-pure HCl to obtain $H_3{}^{32}PO_4$ solution of the desired radioactive concentration.

The advantage of the first route of production is that only a small amount of red phosphorus needs to be irradiated to obtain ^{32}P for bone pain palliation application. The second route from natural sulphur provides NCA ^{32}P. However, such high specific activity is not needed for bone pain palliation therapy. In fact, inactive phosphorus needs to be added during the formulation of the radiopharmaceutical. The second route consumes a large irradiation volume in the reactor and requires elaborate radiochemical processing.

4.2.2. ^{89}Sr

Strontium-89 decays with a half-life of 50.53 days, emitting β^- particles with $E_{\beta(max)}$ of 1.497 MeV. There is no γ emission.

Strontium-89 is usually produced via the $^{88}Sr(n,\gamma)^{89}Sr$ reaction. Despite the high natural isotopic abundance of ^{88}Sr (82.3%), a highly enriched target is needed to avoid producing the ^{85}Sr impurity formed by the nuclear reaction, $^{84}Sr(n,\gamma)^{85}Sr$. ^{89}Sr decays by electron capture, emitting high energy gamma photons of 514 keV. The thermal neutron capture cross-section of $^{88}Sr(n,\gamma)^{89}Sr$ reaction is very low (0.058 b), resulting in a low production yield, even when the irradiation is performed in very high flux reactors. The specific activity of ^{89}Sr formed is inadequate if medium flux nuclear reactors are used and the product is unsuitable for clinical application. Nuclear reactors with high thermal neutron flux are required to produce ^{89}Sr by direct (n,γ) activation to yield adequate specific activity for bone pain palliation.

Strontium-89 is typically produced by thermal neutron irradiation of ^{88}Sr targets and commercially used ^{89}Sr thus has a low specific activity of ~3 GBq/g. Irradiation of 1 g of $SrCO_3$ target in the Maria reactor, NCBJ, Świerk, Poland per five to six cycles of 100 h (neutron flux 2.5×10^{14} $n \cdot cm^{-2} \cdot s^{-1}$) produces 4 GBq (~108 mCi) of ^{89}Sr. To avoid ^{85}Sr co-production, ^{88}Sr targets with very high enrichment (>99.9% ^{88}Sr) are used. The requirement to use highly enriched targets, the low yield due to the poor neutron capture cross-section (0.058 b) and the long irradiation time increase the cost of ^{89}Sr production.

High specific activity ^{89}Sr is produced by the reaction $^{89}Y(n,p)^{89}Sr$ using fast neutrons. Yttrium is a mononuclidic element with 100% ^{89}Y and hence a natural target can be used. The cross-section of this nuclear reaction is very low (0.0002044 b), requiring the irradiation of large quantities of ^{89}Y to obtain reasonable amounts of ^{89}Sr with fast neutron flux. Hence, a large irradiation volume in the reactor is blocked for extended periods of time, contributing to the cost of production. Strontium-89 needs to be separated from bulk irradiated material, which will also contain large quantities of radioactive ^{90}Y formed by the activation of the ^{89}Y target ($^{89}Y(n,\gamma)^{90}Y$). The advantage of this route of production is the possibility of obtaining NCA ^{89}Sr.

In 2003, Chuvilin et. al. proposed a new method for the efficient production of NCA ^{89}Sr, which is based on application of the ^{89}Kr produced as a result of ^{235}U fission. Strontium-89 is formed in the nuclear chain $^{89}Se \rightarrow {}^{89}Br \rightarrow {}^{89}Kr \rightarrow {}^{89}Rb \rightarrow {}^{89}Sr$. This method can be realized in a liquid nuclear reactor in which gaseous fission products are isolated and collected; after an appropriate period of time, ^{89}Sr can be separated. The potential high efficiency of ^{89}Sr production by this method is based on the high cross-section (600 b) of ^{235}U fission. The yield of ^{89}Kr in fission products is ~4.5 % [98].

4.2.3. ^{153}Sm

Samarium-153 is a low energy β^- emitter, decaying with a half-life of 46.27 h. It emits β^- particles with an $E_{\beta(max)}$ of 808 keV (17.5%), 705 keV (49.6%) and 635 keV (32.2%). It also emits gamma photons of 103 keV (28.3%) and 70 keV (5.25%). These gamma photons can be used for scintigraphic imaging.

Samarium-153 is produced by irradiating enriched ^{152}Sm in oxide [99] form in a nuclear reactor by means of the $^{152}Sm(n,\gamma)^{153}Sm$ reaction. There is no radiochemical processing except recovering the target and dissolving it in dilute HCl to obtain $^{153}SmCl_3$. The neutron activation cross-section of 206 b and the resonance integral of 2978 b of ^{152}Sm are advantageous and large activity can be prepared even when the irradiation is carried out in medium flux research reactors. The specific activity obtained is high enough for the preparation of bone pain palliation agents.

Although not preferred, natural samarium can also be used for the production of ^{153}Sm and the specific activity of the product formed is adequate for the preparation of radiopharmaceuticals for bone pain palliation. Natural targets will give long lived radionuclidic impurities, such as ^{145}Sm ($T_{1/2}$ 345 days), ^{151}Sm ($T_{1/2}$ 90 years) and ^{155}Eu ($T_{1/2}$ 4.76 years). However, the radionuclidic impurity burden will not be too high to preclude the use of the product for clinical applications [100].

Samarium-153 is an ideal radionuclide for bone pain palliation and ^{153}Sm–EDTMP (quadramet) is an approved product. The medium energy β^- particles of ^{153}Sm have the advantage of low bone marrow toxicity compared to ^{32}P and ^{89}Sr. The gamma abundance is ~35%, which is on the high side and is a disadvantage for a therapeutic radionuclide. The major disadvantage is its short half-life ($T_{1/2}$ 46.28 h), which is a constraint on the widespread use of ^{153}Sm radiopharmaceuticals.

4.2.4. ^{177}Lu

Lutetium-177 decays with a half-life of 6.73 days and β^- emissions of $E_{\beta(max)}$ 498 keV (78.6%), 385 keV (9.1%) and 176 keV (12.2%). There are also two gamma rays of energy 208 keV (11.0%) and 112 keV (6.4%), which are suitable for scintigraphic imaging. The β^- particles emitted by ^{177}Lu have moderate tissue penetration, making the radionuclide suitable for targeting bone metastases without much irradiation of the bone marrow. Undoubtedly, ^{177}Lu is the most successful radionuclide introduced for TRT at the beginning of this millennium and several radiopharmaceuticals have already been developed in the last 20 years and are in clinical use [101]. Many more radiopharmaceuticals based on ^{177}Lu are likely to be developed in the future for targeted therapy.

Lutetium-177 can be produced by direct or indirect routes and the specific activity achievable from both the methods is very high and suitable for most applications [102]. Both the direct and indirect routes have their own advantages, some of which will be elaborated in the following sections.

4.2.4.1. Direct route for production of ^{177}Lu

Lutetium-176 is the target material for the production of ^{177}Lu via the direct route using the $^{176}Lu(n,\gamma)^{177}Lu$ reaction. The reaction cross-section is (σ 2065 b). The natural abundance of ^{176}Lu is 2.6%; however, highly enriched (>80%) ^{176}Lu targets are commercially available at a reasonable cost. By using highly enriched targets (>90%) and irradiation in very high flux research reactors (~1×10^{15} $n \cdot cm^{-2} \cdot s^{-1}$), specific activities of up to 2.59 TBq/mg (70 Ci/mg) can be obtained, resulting in a radioactive atom percentage >60. Even though the theoretical specific activity of ^{177}Lu is very high at 4.033 TBq/mg (109 Ci/mg), ^{177}Lu, with a specific activity of >740 GBq/mg (20 Ci/mg), is found to be adequate for the radiolabelling of peptides and antibodies [101]. Much lower specific activity is sufficient for the preparation of bone pain palliation agents. Hence, even low–medium flux research reactors can produce large quantities of carrier added ^{177}Lu for therapy.

The specific activity of ^{177}Lu that can be attained in a nuclear reactor while irradiating an enriched ^{176}Lu target depends on the time of irradiation in addition to the neutron flux. The target burnup will be very high in high flux research reactors, as the thermal neutron capture cross-section of the ^{176}Lu target is very high (~2300 b). The activity produced attains a maximum after a certain duration of irradiation and will start decreasing thereafter. The time to attain the maximum activity depends on the neutron flux. The higher the neutron flux, the shorter the time required to attain maximum activity [93]. Hence, it is important to theoretically calculate the optimum time of irradiation for the irradiation position used and the flux available.

The direct route of ^{177}Lu production will result in the formation of a small quantity of ^{177m}Lu, which is a long lived ($T_{1/2}$ 160.5 days), metastable radioisotope of lutetium. There will be no increase in the absorbed dose to tissues other than the target organs as ^{177m}Lu and ^{177}Lu have the same chemical, biological and dosimetric properties. The presence of ^{177m}Lu has been considered to be a problem in certain countries because it is necessary to manage the waste coming from the nuclear medicine department. The average level of burden in ^{177}Lu due to ^{177m}Lu is of the order of ~150 nCi per mCi (150 Bq/MBq) or ~ 1.5×10^{-2} % back calculated to end of bombardment (EOB) [102]. These levels of ^{177m}Lu impurity are not a concern in many countries. As the level of the ^{177m}Lu impurity is low, it is difficult to measure it before dispatch of the product. Activity of ^{177m}Lu is assayed by recording the γ ray spectrum of a decayed ^{177}Lu sample and back calculating the amount of activity at the reference time.

4.2.4.2. *Indirect route for production of ^{177}Lu*

The indirect route of production of ^{177}Lu is followed for the preparation of NCA ^{177}Lu [94, 103, 104]. The nuclear reaction for the production of ^{177}Lu is shown below.

$$\beta^- \longrightarrow \longrightarrow \beta^-$$

$$^{176}Yb(n,\gamma)^{177}Yb \longrightarrow {}^{177}Lu \longrightarrow {}^{177}Hf(stable) \quad (3)$$

$$\sigma\ 2.85b \quad T_{1/2} = 1.9\ h \rightarrow \quad T_{1/2} = 6.73d$$

The ^{177}Lu prepared is NCA and hence will have theoretical specific activity of 4.033 TBq/mg (109 Ci/mg). Lutetium and Yb have identical reaction thermodynamics and kinetics to most of the commonly used ligands for radiopharmaceutical preparation, such as DOTA. Hence, ^{176}Yb target material impurity in the ^{177}Lu results in decreased specific activity of the radiopharmaceutical product. The production of ^{177}Lu through this route therefore demands elaborate radiochemical processing, as large quantities of ^{176}Yb enriched target are required to produce sufficient quantities of ^{177}Lu that is suitable for radiopharmaceutical preparations. It is possible to recover the enriched target after irradiation; however, there are considerable practical difficulties. All these demands add to the production cost of ^{177}Lu via the indirect route.

4.2.5. ^{186}Re

Rhenium-186 decays by β^- emission with a half-life of 90.64 h. It emits medium energy β^- particles with an $E_{\beta(\max)}$ of 1069 keV (80%), 932 keV (21.5%) or 581 keV (5.78%). Being a medium energy β^- particle emitting radionuclide, it is adapted to bone pain palliation. It also emits gamma photos of 137 keV (8.6%), which are ideal for scintigraphic imaging.

NCA ^{186}Re can be produced by proton or deuteron bombardment on a ^{186}W target in a cyclotron at high or moderate energies [105]. However, this is not a preferred route for the production of large quantities of ^{186}Re; moreover, very high specific activity preparation is not essential for bone pain palliation therapy.

Radionuclidically pure ^{186}Re can be prepared by (n,γ) activation of isotopically enriched ^{185}Re targets, the natural abundance of which is 37.4%. The high cross-section (112 b thermal, 1739 b resonance integral) of the nuclear reaction ^{185}Re(n,γ)^{186}Re is highly attractive as large scale production is feasible. The presence of ^{187}Re in the target ought to be avoided to prevent the formation of ^{188}Re as a radionuclidic impurity.

Natural rhenium contains two isotopes, ^{185}Re (37.4%) and ^{187}Re (62.6%). Rhenium-186 has a fairly large thermal neutron activation cross-section of 74.8 b and a 294 b resonance integral. Hence, the irradiation of a natural rhenium target in a reactor will yield a mixture of ^{186}Re and ^{188}Re [106]. As the half-life of ^{186}Re ($T_{1/2}$ = 90 h) is much longer than that of ^{188}Re ($T_{1/2}$ = 17 h), the proportion of the two radionuclides formed depends on the time of irradiation and cooling post-irradiation. Shorter irradiation times will make more of ^{188}Re and less of ^{186}Re, but with low specific activity. Longer irradiation (over 7 days) and cooling for 4 days will bring down the ^{188}Re activity to less than 5%.

The specific activity of ^{186}Re is sufficient to make radiopharmaceuticals for bone pain palliation and radiosynovectomy. Being a medium energy β^- emitter, it is a good radionuclide for bone pain palliation applications. The 90 h half-life of this radionuclide provides logistical advantages for transportation of the radiopharmaceuticals to users.

4.2.6. ^{188}Re

Rhenium-188 decays by β^- emission with a half-life of 16.98 h. It emits high energy β^- particles with an $E_{\beta(\max)}$ of 2.12 MeV. It also emits gamma photons with an energy of 155 keV (with an abundance of 14.9%), suitable for scintigraphic imaging.

Despite being a high energy β^- emitter, ^{188}Re has been used extensively for bone pain palliation therapy. This is mainly due to its availability from a ^{188}W/^{188}Re generator [90]. ^{188}Re can also be prepared by neutron activation of highly enriched ^{187}Re. The cross-section (74.8 b thermal, 294 b resonance integral) is relatively large and hence large quantities of ^{188}Re can be prepared using this route. The ^{188}Re produced is carrier added, but the specific activity obtained when irradiation is performed in medium to high flux research reactors is sufficient for the preparation of bone pain palliation agents.

The preferred route of production of ^{188}Re is as the decay product of a ^{188}W/^{188}Re generator. The parent radionuclide ^{188}W needed to make the generator is produced by double neutron capture on an enriched ^{186}W target (^{186}W(n,γ)^{187}W(n,γ)^{188}W, cross-sections 37.9 b and 64 b thermal, respectively, and 294 b and 2760 b resonance integral, respectively). As ^{188}W is produced by double neutron capture, the production yield is a function of the product of the cross-sections ($\sigma_1 \times \sigma_2$) and the practical yields and specific activity of ^{188}W are very low even when medium flux reactors are used for production [107]. Only a few reactors globally are capable of making the high specific activity ^{188}W needed for the preparation of ^{188}W/^{188}Re generators.

The ^{188}Re prepared from the generator is in NCA form and hence suitable for radioimmunotherapy and peptide receptor radionuclide therapy. Rhenium-188 is also used for the preparation of radiopharmaceuticals for other applications, such as radiosynovectomy and treatment of hepatocellular carcinoma. The extensive use of ^{188}Re based bone pain palliation radiopharmaceuticals has also been promoted through IAEA sponsored projects.

4.2.7. ^{117m}Sn

Tin-117m decays with a half-life of 13.6 days and emits conversion electrons of 127 keV and 129 keV. Unlike β^- particles, conversion electrons have discrete energies. The short path length of these electrons spares the bone marrow. Tin-117m also emits 159 keV (86%) gamma rays. This high abundance of gamma rays coupled with the long half-life of the radionuclide mandates restricted movement of patients to avoid radiation exposure to the public.

Tin-117m is produced by elastic scattering reactivation, involving nuclear reaction ^{117}Sn (n,n′γ)^{117m}Sn. These types of nuclear reactions have very poor cross-sections and hence require the use of very high flux (~10^{15} $n \cdot cm^{-2} \cdot s^{-1}$) reactors [108]. A highly enriched target is essential to produce a sufficient quantity of ^{117m}Sn, making it a costly radionuclide. The specific activity of ^{117m}Sn produced is low but adequate for bone pain palliation therapy.

High specific activity ^{117m}Sn can be produced by using an accelerator and exploiting the nuclear reactions $^{nat}Sb(p,2pxn)^{117m}Sn$ and $^{116}Cd(\alpha,3n)^{117m}Sn$. One report describes the production and purification of ^{117m}Sn prepared via the $^{116}Cd(\alpha,3n)$ reaction [109]. The ^{117m}Sn thus prepared was chelated with an aminobenzyl derivative of DOTA and then conjugated with annexin V-128 to produce an agent for the diagnosis and treatment of vulnerable plaque [109].

4.2.8. ^{170}Tm

Thulium-170 decays to stable ^{170}Yb (99.879 %) by emission of β^- particles and a small fraction to stable ^{170}Er (0.131%) by electron capture with a half-life of 128.6 days. It emits β^- particles with maximum energies ($E_{\beta(max)}$) of 0.968 MeV (81.6 %) and 0.884 MeV (18.3%). It also emits 84.3 keV (3.26%) gamma photons, which are suitable for scintigraphic imaging.

Thulium-170 is produced by irradiating a natural thulium oxide target in a nuclear reactor using the nuclear reaction $^{169}Tm(n,\gamma)^{170}Tm$. The high cross-section (107 b thermal, 1629 b resonance integral) coupled with 100% isotopic abundance of the target provides ^{170}Tm with a reasonably high specific activity even when the irradiation is performed in medium flux research reactors.

Relatively long irradiation times are essential due to the long half-life of the radionuclide. The radiochemical processing is simple, and the product formed will be radionuclidically pure. In one published report [79], ^{170}Tm was produced by thermal neutron bombardment of natural Tm_2O_3 (100% ^{169}Tm). In the reported procedure, ^{170}Tm was produced by irradiating ~10 mg of Tm_2O_3 sealed in a quartz sample and placed in an aluminium can at a thermal neutron flux of 6×10^{13} $n \cdot cm^{-2} \cdot s^{-1}$ for 60 days [79]. The radiochemical processing involved dissolving the irradiated target in 1M HCl with gentle warming followed by evaporating the contents to near dryness. The contents left after drying were dissolved in doubly distilled water to obtain $^{170}TmCl_3$, which is useful for radiolabelling phosphonate ligands.

The radionuclidic purity of the ^{170}Tm was determined by gamma ray spectrometry using an HPGe detector coupled to a 4 K multichannel analyser (MCA) system. Eighty-four keV was the major photopeak; two other X ray peaks at 52 keV and 59 keV were observed in the gamma ray spectrum of ^{170}Tm decay. No other gamma emitting radionuclides were observed. The specific activity of ^{170}Tm produced was 6.36 GBq/mg (172 mCi/mg) at the end of bombardment. This product was suitable for making ^{170}Tm bone seeking amine bisphosphonates as well as macroparticles [79, 110].

4.2.9. ^{175}Yb

Ytterbium-175 decays with a half-life of 4.185 days and emits β^- particles with an $E_{\beta(max)}$ of 470 keV (86.5%) and 73 keV (10.2%). It also emits gamma photons with energies of 396 keV (6.5%) and 282 keV (3.1%).

Ytterbium-175 has not been widely investigated as a therapeutic radionuclide. The feasibility and logistics of its production have been reported on [111]. It can be produced with high radionuclidic purity by thermal neutron activation of an enriched ^{174}Yb target, ^{174}Yb(n,γ)^{175}Yb. The activation cross-section is relatively high at 63 b and hence abundant quantities of the radionuclide could be produced without difficulty. The saturation specific activity formed is 24.09 TBq/g (651 Ci/g) when irradiation is carried out at a neutron flux of 1×10^{14} n·cm^{-2}·s^{-1} using an 100% enriched target accounting for ~0.37% of the theoretical specific activity.

Natural ytterbium cannot be used as a target for production, as it will result in lower specific activity as well as the formation of other radionuclides of ytterbium (^{169}Yb, ^{177}Yb) as well as ^{177}Lu. The nuclear reaction ^{168}Yb(n,γ)^{169}Yb has a very high cross-section (3033 thermal and 19684 b resonance integral). However, as the percentage abundance of ^{168}Yb is small at only 0.13% and the half-life of ^{169}Yb is very long at 32 days, the quantity formed will be relatively small. Ytterbium-169 decays by electron capture (100% K capture) followed by the emission of Auger electrons. The main gamma photons emitted during decay are of reasonably low energy, 177 keV (22%) and 197 keV (35.9%). It is therefore preferable to use a highly enriched ^{174}Yb target for the preparation of ^{175}Yb for medical applications. The specific activity of ^{175}Yb produced by irradiating an enriched ^{174}Yb target is sufficient for the preparation of therapeutic radiopharmaceuticals for bone pain palliation.

In a reported procedure, natural ytterbium oxide powder (spectroscopic grade, >99.99% pure) was used as the target for neutron bombardment at a neutron flux of 3×10^{13} n·cm^{-2}·s^{-1} [112]. A flame sealed quartz ampoule containing a weighed quantity of Yb_2O_3 powder was placed in an aluminium can and sealed using a cold welding process. Irradiations were carried out for 5 to 7 days. At the end of the bombardment, the target was cooled for at least 6 h before chemical processing.

Chemical processing was performed by dissolving the irradiated powder in 1M HCl with gentle warming. The resultant solution was evaporated to near dryness and reconstituted in 5 mL of 0.1M HCl and used for radiolabelling studies. The radionuclidic purity was ascertained by γ ray spectroscopy using an HPGe detector coupled to a 4 K MCA system. Spectra were recorded at regular time intervals for a period of ~3 months to study the radionuclide impurities.

By analysing the spectra of ^{175}Yb recorded regularly over a long period of time, it was found that the γ photopeak corresponded only to ^{169}Yb, ^{175}Yb and ^{177}Lu; no other radionuclidic impurities were identified. The γ ray spectrum of a sample recorded after prolonged cooling (~1 month) only showed the photo peaks of ^{169}Yb.

When neutron bombardment for 5 to 7 days was performed at a flux of 3×10^{13} n·cm^{-2}·s^{-1} on a natural Yb target, 900–1200 GBq/g (24–32 Ci/g) of ^{175}Yb was produced. The ^{175}Yb produced had >95% radionuclidic purity. The specific activity obtained was adequate for utilizing the radioisotope to make bone pain palliation agents [112]. The specific activity of the product can be increased to a large extent by irradiation in high flux research reactors.

Ytterbium-175 of high radionuclidic purity with no contamination of ^{169}Yb and ^{177}Lu can be prepared by using an enriched ^{174}Yb target. Five days of irradiation at a neutron flux of 1×10^{14} n·cm^{-2}·s^{-1} with a 95% enriched ^{174}Yb target will produce ^{175}Yb with a specific activity of ~13.32 TBq/g (360 Ci/g) and the product will be free of ^{177}Lu. Highly enriched (98.6%) ^{174}Yb targets are commercially available and contain >0.01% ^{168}Yb and ~0.4% ^{176}Yb. The availability of the enriched targets at a reasonable cost (presently ~US $5/mg) makes this production route quite cost effective.

There are also reports on isolating ^{175}Yb as a by-product during the production of NCA added ^{177}Lu while using a natural ytterbium target [113, 114].

4.2.10. ^{166}Ho

Holmium-166 is a β^- emitting radionuclide that can be produced in large quantities and is useful for bone pain palliation therapy. It has a half-life of 26.83 h. It emits high energy β^- particles with an $E_{\beta(\max)}$ of 1854 keV (50.0%), 1774 keV (48.7%) and emits gamma photons of 81 keV (6.2%), which are useful for imaging studies.

Holmium-166 is produced by direct neutron activation of natural holmium oxide targets. Holmium is mononuclidic with 100% natural abundance for ^{165}Ho. The neutron activation cross-section is high (64.4 b thermal and 680 b resonance integral); hence, relatively large quantities of ^{166}Ho can be prepared with reasonably high specific activity in medium flux reactors [115, 116].

NCA ^{166}Ho can also be produced from a $^{166}Dy/^{166}Ho$ generator [117]. Dysprosium-166 ($T_{1/2}$ 81 h) in moderate yields is prepared by double neutron capture of ^{164}Dy (2650 b thermal) and ^{165}Dy (3600 b thermal, 22 000 b resonance integral).

The specific activity of the ^{166}Ho produced by direct neutron activation is adequate for the preparation of bone seeking radiopharmaceuticals. Holmium-166 can be complexed with amine phosphonate ligands to produce bone seeking radiopharmaceuticals [116]. The β^- energy of ^{166}Ho is high; hence, there will be considerable bone marrow damage. The quantity of radiopharmaceutical administered has to be limited to reduce myelotoxicity. There are reports of the use of ^{166}Ho for bone marrow ablation in the case of haematological malignancies [118].

4.2.11. ^{90}Y

Yttrium-90 decays with a half-life of 64.1 h and emits β^- particles with an $E_{\beta(\max)}$ of 2.282 MeV (99.983%). It is a pure β^- emitter with no γ emission, but there is a very small component of internal pair production, which can be used for PET–CT imaging [119].

Yttrium-90 is preferentially produced as the decay product of ^{90}Sr from a $^{90}Sr/^{90}Y$ generator. The parent radionuclide ^{90}Sr ($T_{1/2}$ 28.78 years) is a fission product formed at a very high yield (~6%) [120]. Large stocks of ^{90}Sr are available in the world. A series of ion exchange methods for separating ^{90}Y from ^{90}Sr has been developed [89]. An electrochemical separation procedure to separate ^{90}Y from ^{90}Sr has also been reported at the Bhabha Atomic Research Centre [121]. A novel extraction paper chromatography method developed by the same group showed that the ^{90}Sr contamination in ^{90}Y is only at ppm levels and hence the product is safe for the preparation of radiopharmaceuticals for human administration [122].

The high energy (2.28 MeV) β^- particles have a range of ~3.9 mm in soft tissues and hence they will induce considerable damage to bone marrow and are not really suitable for metastatic bone pain palliation.

4.2.12. ^{161}Tb

Terbium-161 is a radionuclide with similar characteristics to those of ^{177}Lu and was proposed as an alternative to ^{177}Lu. Terbium decays with a half-life of 6.88 days, emitting β^- particles with an $E_{\beta(\max)}$ of 593 keV (10.0%), 567 keV (10.0%), 518 keV (66.0%) and 461 keV (26.0%). It also emits low energy gamma photons, with energies of 74 keV (9.8%%) and 49 keV (14.8%).

NCA ^{161}Tb can be prepared by the indirect route, with ^{160}Gd(n,γ)^{161}Gd decaying to ^{161}Tb [123]. The production is not very cost effective as the cross-section is very low (0.77 b) and large quantities of highly enriched target are needed to produce reasonable quantities of ^{161}Tb. Natural gadolinium will not be permitted to be irradiated in large quantities as it is a neutron poison (average σ 49 000 b) and can result in flux disturbances in the reactor. Separation of ^{161}Tb from target Gd needs to be carried out with care as Gd^{3+} and Tb^{3+} are neighbouring lanthanides; hence, it can complex with the ligands used for radiopharmaceutical preparation.

The half-life and decay energies of ^{161}Tb are not very different from those of ^{177}Lu and this radionuclide may not offer any dosimetric advantage over ^{177}Lu as long as only beta emissions are considered. The gamma abundance of ^{161}Tb is higher, though of lower energy and not very suited for imaging studies. Although it can be used for bone pain palliation, there are no reports about its use in the preparation of bone seeking radiopharmaceuticals.

Conversion and Auger electron emissions in the decay process of ^{161}Tb add to the therapeutic effect of ^{161}Tb labelled products. Recently in vitro and in vivo studies on ^{161}Tb-PSMA-617 showed better results and dosimetric calculations compared to ^{177}Lu-PSMA-617 [124].

4.2.13. ^{169}Er

Erbium-169 is a β^- emitting radionuclide, decaying with a half-life of 9.4 days. It emits β^- particles with an $E_{\beta(\max)}$ of 351 keV (55%) and 342 (45%). These low energy β^- particles are ideal for bone pain palliation therapy as the damage to the bone marrow is minimal. Erbium-169 is a pure β^- emitter with no gamma emission and hence cannot be used for imaging.

Erbium-169 is prepared by irradiating enriched ^{168}Er in oxide form in a nuclear reactor. The cross-section is moderate (2.68 b thermal) and thus the specific activity of the radionuclide will be relatively low. Natural targets could also be used, but will have radionuclidic impurities of ^{165}Er and ^{171}Er, and the specific activity will be still lower.

Chakravarty et al. reported the production of ^{169}Er by neutron bombardment of isotopically enriched erbium (98.2% in ^{168}Er) targets at a thermal neutron flux of ~8×10^{13} $n \cdot cm^{-2} \cdot s^{-1}$ for 21 days in a nuclear reactor [125]. A two cycle electrochemical separation procedure was adopted to remove ^{169}Yb co-produced during irradiation. A thorough optimization of irradiation parameters, including neutron flux, irradiation time and target cooling period, was reported. Erbium-169 with a specific activity of ~370 MBq/mg (10 mCi/mg) and radionuclidic purity >99.99% could be obtained. The authors used purified ^{169}Er for complexation with DOTMP, which resulted in a stable complex with radiochemical purity >99%.

Erbium-169 DOTMP has the potential to be developed as a radiopharmaceutical for bone pain palliation.

A recent publication reported the preparation of very high specific activity ^{169}Er by mass separation of the reactor irradiated target to isolate ^{169}Er, resulting in specific activities up to ~240 GBq/mg (6.4 Ci/mg) [126]. However, logistically this mode of production is expensive for large scale preparation of ^{169}Er. Additionally, such high specific activity is not needed for the preparation of bone pain palliation radiopharmaceuticals.

4.3. PRODUCTION OF ALPHA PARTICLE EMITTING RADIONUCLIDES

Not many α particle emitting radionuclides are used for bone pain palliation. Radium-223 chloride ($^{223}RaCl_2$) is an approved radiopharmaceutical for metastatic bone pain palliation [127]. Actinium-225, another alpha emitting radionuclide formulated in the form of ^{225}Ac–PSMA inhibitor ligand, is an efficacious radiopharmaceutical for the treatment of primary as well as metastatic prostate cancer [65]. Both of these radionuclides emit four α particles due to successive decays and all of these particles are useful for therapy. However, if the half-life of the decay progeny is very long, there is a possibility of it leaching out of the bone and redistributing. Alpha particle emitters will play a significant role in TRT in the future due to their ability to deposit the radiation energy specifically to the cancer cells while sparing normal cells [128, 129].

4.3.1. ^{225}Ac

Actinium-225 decays to ^{213}Bi with a half-life of 10.0 days and during this process it emits three alpha particles (Fig. 8 (a)). Bismuth-213 is also an alpha emitter with a half-life of 46 min and it decays through two branching decay modes, with each of these modes emitting one alpha particle (Fig. 8 (b)). Hence, there are four alpha particle emissions in total when an ^{225}Ac atom decays. Actinium-225 labelled radiopharmaceuticals are very attractive, as only very low levels of radioactivity are required due to its long half-life and multiple alpha emissions.

4.3.1.1. ^{225}Ac preparation from long lived parents

Actinium-225 is produced either by isolation from its long lived parent (^{229}Th) or by irradiation of ^{226}Ra in a medium energy cyclotron. Other methods,

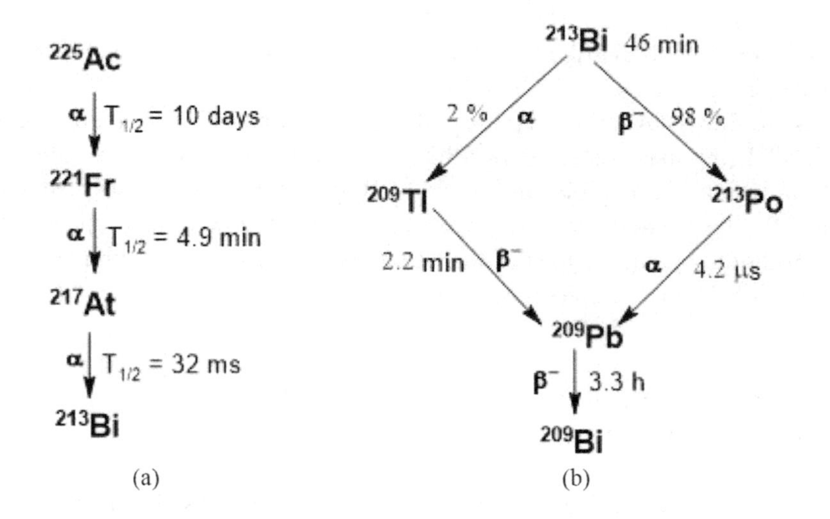

FIG. 8. Decay scheme of ^{225}Ac to ^{213}Bi (a). Bismuth-213 has a half-life of 46 min and decays via two branching decay chains to ^{209}Pb, which decays to stable ^{209}Bi (b) (courtesy of T. Das, Bhabha Atomic Research Centre).

such as a photonuclear reaction and spallation of ^{232}Th using high energy protons, are also being explored.

Thorium-229 belongs to the artificially produced 4n + 1 chain of the ^{237}Np series (Fig. 9). The parent radionuclides, ^{237}Np ($T_{1/2}$ 2.14 × 10^6 years) and ^{233}U ($T_{1/2}$ 1.6 × 10^5 years) are not naturally occurring and are the products of the weapons programme. Both of these radioisotopes have very long half-lives; hence ^{229}Th can only be isolated once, as further growth will be minimal. Thorium-229 ($T_{1/2}$ 7880 years) stock is isolated in the United States of America as well as in the Russian Federation from existing ^{237}Np or ^{233}U stocks. Thorium-229 decays to ^{225}Ra ($T_{1/2}$ 14.0 days), which is the parent radionuclide for the separation of ^{225}Ac. Overall, the accepted global annual production of ^{225}Ac from ^{229}Th is 63 GBq (1.7 Ci) [130].

4.3.1.2. Cyclotron production of ^{225}Ac

Actinium-225 is produced by irradiating ^{226}Ra with a proton beam in a cyclotron by nuclear reaction, ^{226}Ra(p,2n)^{225}Ac; the cross-section is 700 mb, peaking at 15 MeV (Fig. 10) [85]. Radium-226 is available in nature, but it is radioactive and hence target preparation, irradiation, post-irradiation handling and recovery require additional facilities and skills. However, this route is highly promising, as large quantities of ^{225}Ac can be prepared using this method.

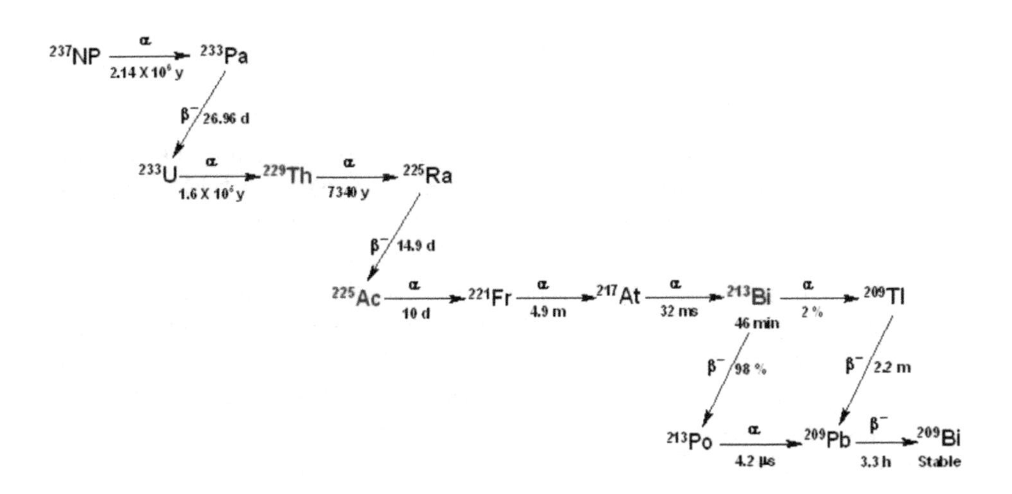

FIG. 9. Decay of the artificially produced $^{237}Np/^{233}U$ series. Radium-225, ^{225}Ac and ^{213}Bi are the alpha emitting radionuclides suitable for therapy in this series (courtesy of T. Das, Bhabha Atomic Research Centre).

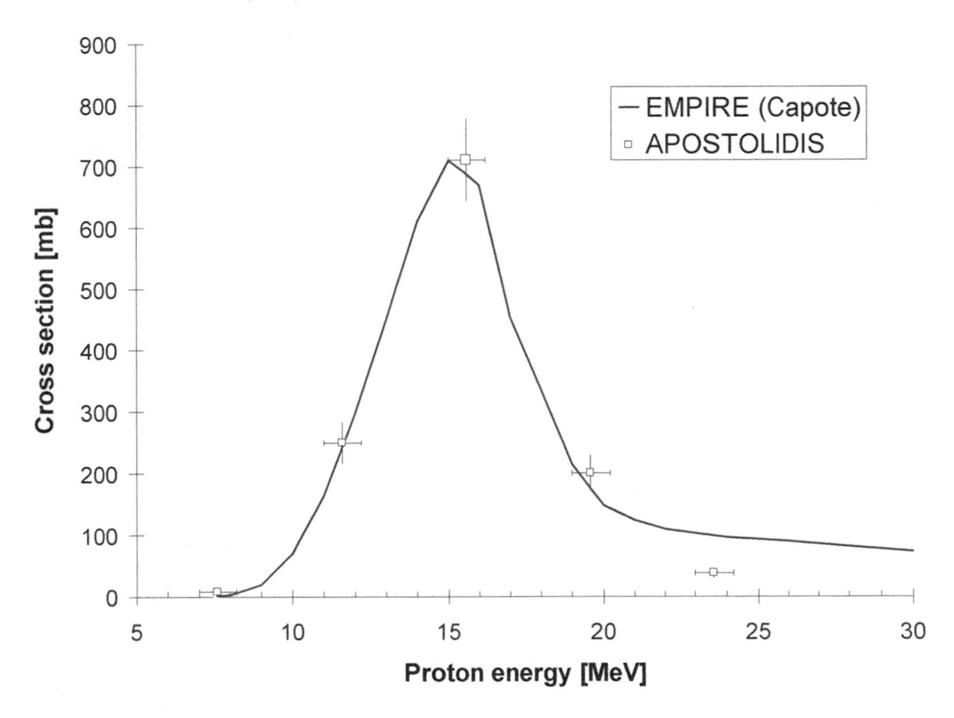

FIG. 10. Excitation function of $^{226}Ra(p,2n)^{225}Ac$.

4.3.1.3. *Production of ^{225}Ac using photonuclear reactions*

A few laboratories, such as Argonne National Laboratory, USA, and TRIUMF, Canada, have reported production of ^{225}Ac using the photonuclear reaction of $^{226}Ra(\gamma,n)^{225}Ra$ [130]. The reaction threshold is 6.4 MeV and the maximum cross-section of the reaction is 532 mb at an energy of 13.75 MeV. The ^{225}Ra produced decays with a half-life of 14.9 days to ^{225}Ac. A company in the USA involved in radioisotope production is using superconducting electron linear accelerators to produce ^{225}Ac and other alpha emitters [131]. In their method, ^{225}Ac produced from a nitrate based solution of ^{226}Ra is eluted continuously from the target vessel and centrifugal contactors are used to harvest and purify ^{225}Ac through a separation cascade. This method has the advantage that it does not generate any long lived ^{227}Ac contamination in the product.

4.3.1.4. *Spallation of ^{232}Th*

Spallation of ^{232}Th in high energy proton accelerators for ^{225}Ac production is being pursued by a few facilities in the USA and Canada. Thorium-232 has several positive attributes, such as high abundance, suitable thermochemical properties and a high spallation cross-section, and it is inexpensive and an easy target to handle. Spallation of ^{232}Th results in a significant yield of ^{225}Ac as well as ^{225}Ra, which decays to ^{225}Ac. Long lived ^{227}Ac contamination is a concern from the clinical safety point of view, while the effects of following this route are not yet fully understood.

4.3.2. ^{223}Rn

Radionuclide ^{235}U is naturally occurring. During its decay process several radionuclides are produced, including ^{223}Rn ($T_{1/2}$ 11.43 days), which produces three alpha particles during its decay to ^{211}Pb ($T_{1/2}$ 36.1 min). Figure 11 shows the decay pathway of ^{223}Rn to ^{211}Pb ($T_{1/2}$ 36 min) and to ^{211}Bi ($T_{1/2}$ 2.14 min), with two subsequent independent decay modes, with α decay (99.7%) resulting in ^{207}Tl and β^- decay resulting in ^{211}Po; both of these radionuclides decay further to stable ^{207}Pb.

Naturally occurring ^{235}U will have sufficient quantities of ^{227}Ac ($T_{1/2}$21.7 years), which can be isolated and kept as a source for the preparation of ^{223}Ra. Actinium-227 decays to ^{227}Th by β^- decay, which in turn decays to ^{223}Ra by α decay (Fig. 12).

Among the various decay products of ^{223}Ra, ^{211}Pb ($T_{1/2}$ 36.1 min) poses the potential problem of migration from the target (redistribution). Radon-219 is formed as a gas, but, being short lived and a bone seeker, it is expected to remain

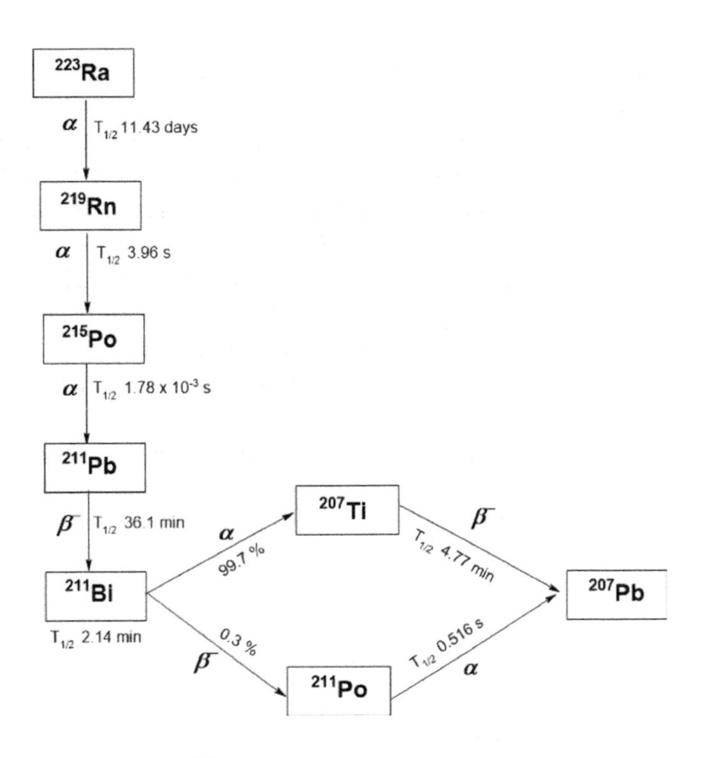

FIG. 11. Decay scheme of ^{223}Ra (courtesy of T. Das, Bhabha Atomic Research Centre).

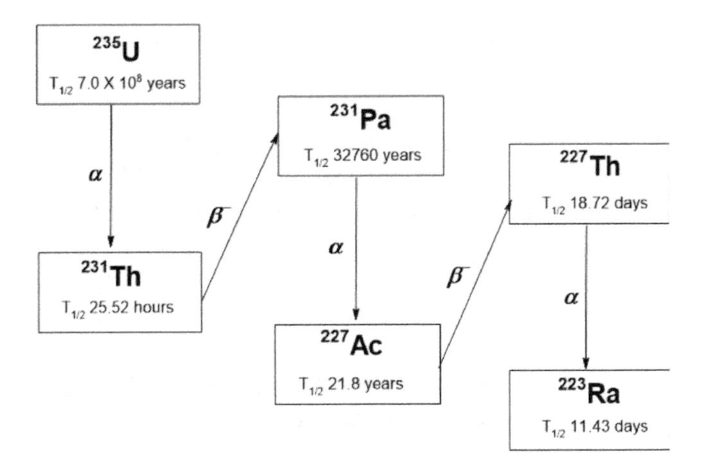

FIG. 12. Decay scheme for ^{235}U, leading to the production of ^{223}Ra (courtesy of T. Das Bhabha Atomic Research Centre).

in the bone and not leach out. The rest of the decay products are also expected to be contained within the target, as their half-lives are too short.

Typically, ^{223}Ra is obtained from an $^{227}Ac/^{223}Ra$ generator. Actinium-227 ($T_{1/2}$ = 21.8 years) decays predominantly via ^{227}Th ($T_{1/2}$ = 18.7 days) to ^{223}Ra. Actinium-227 can be prepared by thermal neutron irradiation of ^{226}Ra in a nuclear reactor through the $^{226}Ra(n,\gamma)^{227}Ra \rightarrow ^{227}Ac$ reaction. Usually, ^{226}Ra is irradiated with epithermal neutrons using a $RaCO_3$ target [132]. The effective cross-section value of the ^{227}Ac production is 14 ± 4 b. After irradiation, the target is processed using radiochemical separation techniques. The ^{227}Ac is separated from the remaining ^{226}Ra, purified and stockpiled. The ^{227}Ac obtained is used for production of $^{227}Ac/^{223}Ra$ generator.

Usually, for immobilization of ^{227}Ac, an extraction chromatographic column containing P,P′-di(2-ethylhexyl) methane diphosphonic acid immobilized on silica (Dipex-2) is used [133]. In this procedure ^{223}Ra is separated from the parent radionuclides $^{227}Ac/^{227}Th$ by elution with 1M HCl. The next step is based on the concentration and final purification of ^{223}Ra by its sorption on the cation exchange resin (AG-50W-X12 resin) and subsequent elution with 8M HNO_3. The solution is evaporated and ^{223}Ra residue is dissolved in sodium chloride/sodium citrate solution.

An alternative approach for the production of ^{225}Ac involving a cyclotron method was reported in 2017 [134]. In this production process, another actinium radionuclide, ^{227}Ac, is formed as a 1% impurity during high energy proton irradiation of ^{232}Th. After the decay of ^{225}Ac ($T_{1/2}$ = 10 days), the remaining ^{227}Ac can be used to produce ^{223}Ra [134].

Radium-223, in the form of $^{223}RaCl_2$ (alpharadin; Xofigo, USA), has undergone phase III clinical trials in symptomatic prostate cancer patients with bone metastasis [127]. The radiopharmaceutical (as $^{223}RaCl_2$) was administered by intravenous injection once a month for 4 to 6 months in the above clinical trials, and showed effective pain relief.

5. PRE-CLINICAL EVALUATION OF BONE PAIN PALLIATION AGENTS

5.1. INTRODUCTION

Bone is a common site for metastasis of several types of cancers and is the major source of pain in cancer patients. Several mechanisms have been reported for occurrences of bone metastasis in different types of cancers involving different cellular and molecular mediators [135].

The currently available radiopharmaceuticals used for bone pain palliation can broadly be classified into two major categories based on their chemical structure. Some of these agents are simple inorganic molecules containing one radioactive element, such as $^{89}SrCl_2$, $^{223}RaCl_2$ and $H_3{}^{32}PO_4$, while others are radiolabelled organic (phosphonic acid) moieties, such as ^{177}Lu/^{153}Sm–EDTMP, ^{177}Lu/^{153}Sm–DOTMP and ^{186}Re/^{188}Re–HEDP (hydroxyethylidene diphosphonic acid) [136, 137]. The agents based on simple inorganic salts act as bulking agents, which are uniformly deposited in the bone matrix and become part of the bone. These agents have high affinity (HA) for bone mineral, as they act as an analogue of the calcium present in the bone matrix. On the other hand, radiolabelled phosphonic acids are usually adsorbed on the surface of the HA particles [138]. A chemisorption process or oxide formation has been postulated as the probable mechanism for the target specificity of such agents. It has been reported that osteoblasts and osteoclasts play a major role in bone remodelling. Osteoblast activity is generally seen in metastatic prostate cancer, while osteoclast activity is usually associated with breast, lung and kidney metastases [139, 140]. Osteoblast or osteoclast activities enhance the availability of calcium phosphates and HA particles at the metastatic sites, enhancing the uptake of bone seeking radiopharmaceuticals in skeletal metastases [141].

It is important to understand the mechanism of uptake of any new radiopharmaceutical developed for bone pain palliation. Hence, any new radiolabelled agent that has been developed for bone pain palliation needs to undergo systematic and elaborate pre-clinical evaluation before regulatory bodies can be approached to obtain permission for the initiation of clinical evaluation of the agent in patients. These pre-clinical studies are broadly classified into two categories, namely in vitro studies and in vivo studies.

5.2. IN VITRO CELLULAR STUDIES

Some of the in vitro studies, usually carried out to ascertain the potential of a bone pain palliating radiopharmaceutical, are described in the following sections.

5.2.1. Binding efficiency of bone seeking radiopharmaceuticals towards bone matrix

Studying the binding efficiency of bone seeking radiopharmaceuticals towards bone matrices is a good way to determine the effectiveness of any bone pain palliation agent. HA particles, which are the basic chemical constituent of the bone matrix, usually serve as the primary material for carrying out binding studies [142]. However, these particles cannot mimic the in vivo conditions exactly, as the bone cells actually consist of several types of proteins, lipids, growth factors and vitamins, in addition to the HA particles. Thus, there is a need to study the uptake of these agents in a dynamic state of bone mineralization, which helps to mimic the bone matrix more accurately. Bone matrix is usually generated in vitro by culturing well established osteosarcoma cell lines using specific protocols for mineralization.

5.2.1.1. Preparation of osteosarcoma cell lines

In order to prepare bone mineralization, either MG63 or Saos2 cell lines are used. These cell lines are derived from primary osteosarcoma and represent human osteoblast like cells and a source of bone related molecules [143]. Minimum essential media supplemented with 10% foetal calf serum (Gibco), 2 mM L-glutamine and antibiotic/antimycotic solution are used for culturing and cell lines are maintained at 37°C in a humidified 5% CO_2 atmosphere; 10 nM dexamethasone, 10 mM β-glycero-phosphate and 50 μg/mL ascorbic acid are used as osteogenic supplements in minimum essential media to induce bone mineralization in MG63 and Saos2 cell lines [144].

Cells are harvested and plated in six well plates (~1×10^4 cells of either MG63 or Saos2 per well) and incubated overnight in complete media, and are then cultured in osteogenic media containing 10% foetal calf serum (Gibco), 2 mM L-glutamine and antibiotic/antimycotic solution. Every second or third day fresh media need to be added carefully to avoid dislodging cells; this procedure is continued for 3 to 4 weeks (minimum 21 days). The extent of bone formation can be checked by following the standard procedures reported in the literature (i.e. by staining with alizarin red) [145]. The intensity of the red colour, which can be seen by the naked eye, indicates the extent of bone formation in the media. Mineralized bone cells can be stored at −80°C to carry out cell binding studies in the future or may be used immediately for experiments.

5.2.1.2. Binding of bone pain palliating agents to the mineralized bone

In order to study the extent of binding of bone seeking radiotracers to the mineralized bone, equal amounts of mineralized bone, as prepared by the above mentioned procedure, are taken in several tubes to which different amounts of the tracer are added. The agent is incubated with the mineralized bone for 2 h, after which the mineralized bone is washed with phosphate buffered saline. The washing is repeated thrice and the radioactivity associated with the mineralized bone cells is measured and compared with the total activity added per tube.

The percentage binding of the radiolabelled agent with the mineralized bone is calculated from the total activity added per tube to determine the percentage binding of the radiolabelled agent with the mineralized bone.

5.2.2. Studies to determine the cellular toxicity of bone seeking radiopharmaceuticals

Radiation plays a major role in the induction of cell toxicity and apoptosis, imparting direct and indirect damage to the cells. DNA strand breaks are the major manifestation of direct radiation damage [146]. Indirect damage from radiation mostly occurs through the radiolysis of water molecules inside and near the cells. Different free radical species formed after radiolysis of water cause secondary damage to DNA, protein and lipid molecules. Cell membranes, which consist of lipid and protein, undergo peroxidation by the free radicals generated by radiation; this is known to be the primary cause of loss of membrane integrity [147]. Once the membrane loses its permeability, intracellular constituents of the cells are released from the cell. Therefore, studying the cell membrane integrity provides a good measure of the extent of cell damage caused by radiation.

In order to study membrane integrity, cells are stained with dyes that are not taken up by the living cells, but rather are internalized when the membrane of the cell is ruptured. Dyes such as trypan blue, 7-aminoactinomycin D (7-AAD) and propidium iodide (PI) are used for this purpose and the number of dead and live cells can be counted by using either a microscope or a flow cytometer.

Another way to estimate the cell membrane damage or cell toxicity is to determine the extent of the release of intracellular constituents. This can be accomplished by using lactate dehydrogenase (LDH) as a marker, since LDH is only released from the cells when the cell membrane is damaged. Being an enzyme, it can easily be estimated by measuring the enzyme activity [148, 149]. Cellular toxicity studies can be performed in bone osteosarcoma cell lines by incubating the cells with bone pain palliation agents for certain pre-designated time periods and carrying out various assays, which are briefly described below.

5.2.2.1. Trypan blue dye uptake to study cell membrane integrity

Bone osteosarcoma cell lines (MG63 or Saos2) are grown in MEM supplemented with 10% foetal bovine serum and antibiotic/antimycotic solutions following the protocol mentioned above. Cells are treated with different quantities of bone pain palliating radiopharmaceuticals and incubated for various time periods. Subsequently, cells are harvested after trypsinization and a trypan blue dye uptake study is carried out following the protocol described in the literature [144]. In brief, the treated cells are incubated with 0.4% trypan blue dye for 5 min and counted using a bright field microscope with the help of a haemocytometer to determine the percentage viability or percentage death of the cells. Dead cells take up the dye and thus appear dark under the microscope, whereas viable cells remain unaffected.

5.2.2.2. LDH assay

Radiation damaged cells release intracellular LDH, which can be estimated by measuring the enzyme activity. For carrying out the LDH assay, the supernatant from cell harvesting media is collected and centrifuged at 2000 rev./min to remove the floating dead cells. Subsequently, cell supernatant is added in multi-well plates and the LDH assay is carried out following the kit manufacturer's protocol. In brief, the constituents of the LDH assay kit (Sigma-Aldrich), namely LDH assay substrate, cofactor and dye, are mixed together by taking them in equal proportion to make the LDH reaction mixture as described in the kit protocol. This reaction mixture is added in twice the volume of culture media present in the multi-well plate, mixed and incubated in the dark at room temperature for half an hour. The enzyme substrate reaction is stopped by the addition of 1N HCl. A pinkish red colour develops, which is measured at 490 nm using a UV–Vis spectrophotometer. The percentage release of LDH, which is a measure of the extent of cell damage, is calculated with reference to the control sample following the procedure described in the literature [144].

5.2.2.3. Study of cell proliferation by MTT assay

Cells exposed to the radiation from bone pain palliating radiopharmaceuticals may lose their cell division potential and ultimately die. Live cells reduce MTT (3-(4,5-dimethylthiazol-2-yl)-2,5-diphenyltetrazolium bromide) dye into formazan, while dead cells have no effect. For carrying out the MTT assay, treated and control cells are harvested after trypsinization and counted using a haemocytometer. Approximately 1×10^4 cells/well are plated on a multi-well plate and the MTT solution is added to this. Cells are subsequently incubated for 4 h and the formazan formed during the reaction is solubilized in either 20% SDS (sodium dodecyl sulphate) in 50% DMF (dimethyl formamide) or in DMSO (dimethyl sulphoxide). This generates the purple–blue coloration, which is measured at 570 nm with reference to 630 nm to calculate the percentage of cell proliferation [144, 150].

5.2.2.4. Study of cell cycle

Radiation has distinct impacts on the cell cycle and thus studying the phases in radiation exposed cells provides valuable information regarding the effectiveness of any radiopharmaceutical. In vitro cultured cells have a typical pattern of population distributions in G1, S and G2/M phases. Cells that have encountered radiation exhibit reshuffling of population distributions among different phases, which is attributed to the differences in radiosensitivity among various cell cycle phases. Osteosarcoma cells (MG63 or Saos2) are synchronized and treated with different concentrations of a bone pain palliating radiopharmaceutical and separately with equivalent amounts of unlabelled ligand, which is used as the control vehicle. Cells are harvested and incubated in 70% ethanol overnight at 4°C to fix them. Subsequently, the cells are washed with phosphate buffered saline to remove any traces of alcohol and mixed with cell cycle reagents (PI dyes supplemented with RNase) or any commercial cell cycle reagent for flow cytometry. Samples are acquired and analysed using a flow cytometer, which provides valuable information about the population distribution of the cells among the various phases [151].

5.2.3. Study of apoptotic cell death

Cells that have encountered radiation often show DNA strand breaks. If cells are unable to repair the DNA damage, they opt for programmed cell death, which is known as 'apoptosis'. Therefore, various studies related to apoptotic cell death provide relevant information regarding the effectiveness of a radiopharmaceutical in destroying the cancer cells. The protocols for some such studies, which are usually employed to evaluate the potential of bone seeking radiopharmaceuticals, are briefly mentioned below.

5.2.3.1. Estimation of apoptotic DNA fragmentation

Apoptosis is characterized by a typical DNA cleavage, which yields DNA fragments of the order of 7 kDa. Fragmented DNA can be estimated to obtain an idea of the magnitude of apoptosis [152]. Apoptosis is estimated using the well known apoptotic DNA ladder assay kit and the ELISA (enzyme-linked immunosorbent assay) based assay for poly(ADP–ribose)polymerase (PARP) cleavage and in situ cell death detection kit. The protocol described in the commercially available in situ cell death detection ELISA kits is used to estimate DNA fragmentation. Briefly, $\sim 1 \times 10^5$ cells of MG63 that have been treated with a bone pain palliating radiopharmaceutical are lysed and the supernatant is carefully transferred to new tubes. Anti-histone antibody is added to a multi-well plate and incubated overnight at 4°C. Subsequently, cell lysates are added in each well and incubated for 2 h. After completion of the incubation period, the wells are washed with buffer (provided with the kit) and the cells are further incubated for a period of 2 h after the addition of anti-DNA horseradish peroxidase to the wells. The wells are washed again and incubated with substrate solution for 20 min. The colour developed is quantified at 405 nm in an ELISA reader. Estimation of DNA fragmentation is usually expressed as an enrichment factor, which is the ratio of the optical density of the treated sample to that of the control sample [144].

5.2.3.2. Study of apoptotic cell death by flow cytometry

Flow cytometry plays an important role in estimating the absolute count of cells, which are in different stages of apoptosis. Flow cytometry estimates early or late apoptotic cells as well as live cells and cell debris, which are formed due to the interaction of cells with the radiation emitted from bone pain palliating radiopharmaceuticals. Bone osteosarcoma cells are treated with different amounts of bone pain palliating agent for specified time periods. Subsequently, cells are harvested and resuspended in 1% BSA (bovine serum albumin) containing phosphate buffered saline. Apoptotic detection reagent is mixed with the cells and incubated for 20 min at room temperature in a dark environment. Samples are acquired and analysed by using a flow cytometer, which provides information about the number of cells present at various stages of apoptosis [151].

5.2.3.3. Study of protein expression related to apoptotic cell death

Apoptosis is characterized by the expression of several proteins. These proteins are quantified using the Western blotting technique. Osteosarcoma cells (MG63 or Saos2) are treated with a bone pain palliating radiopharmaceutical for a predefined period of time and harvested. Cells are lysed in cell lysis buffer (10 mM Tris pH7.4, 100 mM NaCl, 1 mM EDTA, 1 mM NaF, 20 mM $Na_4P_2O_7$, 20 mM Na_3VO_4, 1% Triton X-100, 10% glycerol, 0.1% SDS, 0.5% deoxycholate and 1 mM PMSF) spiked with protease inhibitor. The protein concentration in the sample is estimated and the protein is loaded on SDS PAGE (sodium dodecyl sulphate polyacrylamide gel electrophoresis) for gel electrophoresis. The protein is transferred onto a nitrocellulose membrane by electroblotting and 5% non-fat milk protein is used to block the non-specific sites on the nitrocellulose membrane. The membrane is incubated with primary antibodies for PARP, such as p38 and phosphor p38, and β actin, followed by treatment with a second antibody. The nitrocellulose membrane is incubated with ECL reagents and exposed to X ray film. Subsequently, the X ray film is processed and developed. Protein band densitometry is performed to determine the expression of proteins [153].

5.3. IN VIVO PRE-CLINICAL STUDIES

In vivo pre-clinical studies performed with bone pain palliating radiopharmaceuticals usually comprise pharmacokinetic evaluation and biological distribution of the agent in normal as well as diseased animal models. Brief details concerning the procedures for carrying out in vivo evaluations in a small animal model are provided below.

5.3.1. Pharmacokinetic evaluation and biodistribution study

Institutional animal ethics committee approval is mandatory for all experiments involving animals. Pharmacokinetic evaluations and studies related to the biological distribution of bone pain palliating radiopharmaceuticals are carried out in animal models such as Swiss mice or Wistar rats. The animals are divided into multiple groups, with each group containing three to five animals for every time point to be studied. The duration of the experiment depends on the half-life of the radionuclide used for the preparation of the bone pain palliation agent. Each animal is administered with 1.85–3.70 MBq (50–100 μCi) of the bone pain palliating radiopharmaceutical, usually through one of the lateral tail veins of the animal. Animals are sacrificed after the designated time period and the major organs/tissues are excised. Blood is collected by cardiac puncture immediately after sacrificing the animal. The organs are washed with normal saline, dried and weighed using an analytical balance. The radioactivity associated with each organ is measured by using a suitable radioactive counter, usually a NaI(Tl) scintillation detector. The activity associated with each organ is calculated and usually expressed as the percentage of injected activity per organ (%IA/organ) or the percentage of injected activity per gram (%IA/g) of the organ. Depending on the objective of the study — variation of the vector or variation of the radioactivity as a function of time — a decay correction may have to be implemented. The accumulation of radioactivity in the skeleton, blood and muscle is usually determined by considering that these organs/tissues constitute 10%, 7% and 40% of the total body weight of the animal, respectively [59, 154].

5.3.2. Animal models for bone metastasis

An initial approach to developing an animal model to study metastatic bone pain was carried out by Arguello et al. [155]. To develop bone metastases in animals, B16 melanoma cells were injected in the left cardiac ventricle of the mice. These cells travelled to various organs, including bone marrow of the skeletal system, resulting in the formation of melanotic tumour colonies. Tumour colony proliferation in the intramedullary space of the bone marrow resulted in destruction of surrounding bone tissues, which is similar to clinical observations for tumour metastases in breast, prostate and lung cancers.

However, the model above did not have any control over other organ metastases [156]. Thus, there was a need for a better animal model to study the efficacy of metastatic bone pain palliating agents. To overcome this, instead of injecting the tumour cells into the left ventricle of the heart, tumour cells were injected directly into the intermedullary space of bone marrow in the tibia and femur. Schwei et al. described comprehensive studies in a mouse model where fibrosarcoma cell lines were injected directly into the medullary cavity of the distal femur by performing left knee arthrotomy on anesthetized mice [157]. This method is superior to the previous one as it leads to controlled metastases in the bone without invading adjacent soft tissues. It is worth mentioning that rats offer various advantages over mice for carrying out animal studies related to bone pain [158]. Medhurst et al. described the first rat model to study bone pain in which the syngeneic MRMT1 mammary gland carcinoma cell line was injected into the tibia of female Sprague Dawley rats by performing arthrotomy on anesthetized rats [159].

5.3.2.1. Developing animal models for bone metastasis protocol - 1

The most simplistic of all experimental models of ectopic bone formation is subcutaneous implantation [160]. In general, in order to obtain an ectopic bone disc, a nano-crystalline HA scaffold is necessary, such as demineralized bone matrix particles (DBMPs), to favour in vivo bone formation. DBMPs are prepared from diaphysis (without muscular and medullar tissue), obtained from the bones at the front and rear extremities of rats. Diaphysis is cut into pieces and washed with distilled water for 2 h, dried with absolute ethanol for 1 h and with ethyl ether for 30 min, and then crushed by hammering until micrometric particles are obtained. By using sieves, the DBMPs (with a size between 90 and 425 μm) are selected and demineralized with a 0.6M HCl solution for 2 h. Subsequently, the DBMPs are washed with plenty of water until the pH is neutralized and then dried using the process described above.

For subcutaneous implantation of DBMPs, anaesthetized Wistar rats are used; two aseptic incisions (approximately 1 cm long) are induced on the dorsum of the rodent. With the help of a spatula, the skin is separated from the muscle to form a kind of pocket, where 15 mg of DBMPs are deposited. The incision is sutured with 3–0 silk thread and the day of the implant is designated day 0. Bone disc evolution is monitored periodically, and on day 16, the bone discs are retrieved. The retrieved bone disc is kept in static cultures in Dulbecco's Modified Eagle Medium (DMEM) 10% foetal bovine serum. The media are changed every 48 h. On day 2, 1×10^5 LNCaP cells are used for inoculation and the discs are maintained for further growth for two days. Bone discs with sufficient tumour cells are implanted subcutaneously into the left and right flanks of 10 week old male Nu/Nu mice under anaesthesia. Cellular uptake of the target specific radiopharmaceutical in the implanted bone metastasis discs is monitored on day 7, post-implantation, with a pre-clinical imaging device.

5.3.2.2. Developing animal models for bone metastasis protocol - 2

Tumour cells of rat mammary gland carcinoma are used for developing animal models with bone metastasis [161]. The right legs of anaesthetized rats are shaved and the skin is disinfected. Superficial incisions are made in the skin. A 22 gauge needle is used to drill a hole at the site of the intercondylar eminence of the tibia and a suspension of tumour cells (1×10^8 cells/mL) is injected. Thereafter, gelatine sponge dissolved in sterile water is used to seal the drilled hole. The injection site is closed using bone wax and dusted with penicillin powder. Animals are allowed to recover from the surgery for 3 days prior to radiopharmaceutical administration and rats are scanned with a pre-clinical imaging device.

5.4. CONCLUSION

In vitro and in vivo evaluation of potential radiotracers is an important area of radiopharmaceutical research. In the past, biodistribution studies were performed to ensure that developed radiotracers had the right pharmacokinetic characteristics to be used as organ seeking diagnostic agents. The data collected from such studies were also used to assess the dosimetry. Diagnostic radiopharmaceuticals were employed for clinical use in volunteers based on those results. It was not uncommon in the early days for radiochemists who had developed products or nuclear medicine physicians to volunteer to be the first recipients of new diagnostic tracers. However, the evaluation of therapeutic radiopharmaceuticals posed several challenges in terms of determining the safety and efficacy of new tracers. The IAEA has addressed this issue through one of its CRPs, Comparative Evaluation of Therapeutic Radiopharmaceuticals, organized from 2002 to 2005. Several assays and protocols were developed during this programme and published [162]. The protocols described in this section are an extension of the same, with a more specific approach to the development of bone pain palliation agents.

6. DOSIMETRY CALCULATION FOR ADMINISTERED RADIOPHARMACEUTICALS

6.1. DOSIMETRY

Radiation therapy involves the use of ionizing radiation for the treatment of cancer and can be administered either from external radiation sources (external beam radiotherapy, EBRT) or through a radiopharmaceutical that concentrates selectively in cancer cells (TRT or molecular radiotherapy, MRT). The constraint of radiotherapy (both EBRT and TRT) is to maximize the radiation being delivered to the cancer tissues while at the same time keeping the radiation being delivered to neighbouring healthy tissues and organs within acceptable limits. Dosimetry aims to assess the absorbed dose delivered to biological tissues by ionizing radiation. Characterizing the radiation delivered to both clinical targets and critical organs is a means of monitoring and optimizing therapy. In that respect, dosimetry is an essential component of radiotherapy.

6.2. DOSIMETRY IN TARGETED RADIONUCLIDE THERAPY

The principle of TRT is to concentrate short range radiation in the neighbourhood of a biological target via a molecular vector. The vector ensures the selectivity of the therapy, and the short range of radiation emitted by the radionuclide ensures that only targeted volumes are irradiated and surrounding tissues are spared.

The radionuclides proposed for TRT therefore mostly emit short range alpha particles, beta particles or Auger electrons [163]. The dosimetric properties of these particles depend on:

(a) The type of particles emitted;
(b) The energy of the emitted particles;
(c) The half-life of the radionuclide;
(d) The possible presence of unstable decays (mostly in the case of alpha emitters).

In the case of bone pain palliation agents, the radiopharmaceutical is delivered to the metastatic bone tissues. Bone metastases are a highly heterogeneous medium even at the microscopic scale. Further, the proximity of bone metastases and red bone marrow, which is a critical organ for radiation, means that studying the energy deposition pattern is essential. It is also a very challenging research domain with implications for the dosimetric approaches that can be implemented [164, 165]. However, the dosimetry formalism used to calculate absorbed doses is the same as that for any other target specific radiopharmaceuticals.

6.3. DOSIMETRIC FORMALISM AND APPLICATION TO SMALL SCALE DOSIMETRY

Absorbed dose calculation requires the determination of the number of radioactive disintegrations taking place as well as the accurate location in all source regions. Absorbed dose calculation also requires the implementation of radiation transport from the emission point (radioactive decay in the source) to the volume where the absorbed dose needs to be calculated (target). This is conventionally presented according to the medical internal radiation dose (MIRD) formalism [166] in the following equations:

$$\bar{D}_{(k \leftarrow h)} = \tilde{A}_h \sum_i n_i E_i \Phi_i (k \leftarrow h) \quad (4)$$

where

$\bar{D}_{(k \leftarrow h)}$ is the average absorbed dose (in Gy) delivered to target k from decays in the source h;
$\tilde{A}_h$ is the cumulated activity (in Bq/s), that is the total number of decays in source h;
n_i is the yield, namely the probability of emission of the ith radiation per decay;
E_i is the energy (in joules) of the ith radiation considered;

and $\Phi_i (k \leftarrow h)$ is the specific absorbed fraction (kg^{-1}) from source h to target k for the ith radiation, that is the ratio of the absorbed fraction $\phi_i (k \leftarrow h)$ /the mass of the target.

The absorbed fraction $\phi_i (k \leftarrow h)$ is the ratio of absorbed vs emitted energy for a given source/target/radiation configuration. $\phi_i (k \leftarrow h)$ varies between 0 and 1.

$$\Phi_i (k \leftarrow h) = \frac{\phi_i (k \leftarrow h)}{m_k} \quad (5)$$

where m_k is the mass of the target.

The cumulated activity is obtained by integrating the time activity curve in the source:

$$\tilde{A}_h = \int_0^\infty A_h(t)\,\mathrm{d}t \tag{6}$$

In the above equation, the integration limits (o and ∞) are generic and can be replaced in practice by the time radioactivity arrives at the source *h* and the time all radioactivity in source *h* has gone.

Grouping all parameters that do not depend on time, one can define the so-called *S* value or *S* factor, $S_{(k\leftarrow h)}$, which represents the average absorbed dose in target *k* from source *h* per decay in source *h* (in Gy·Bq^{-1}·s^{-1}):

$$S_{(k\leftarrow h)} = \sum_i n_i E_i \Phi_i(k \leftarrow h) \tag{7}$$

As n_i and E_i are physical parameters, characteristics of the radionuclide, the hypothesis is that $S_{(k\leftarrow h)}$ does not depend on time. This means that $\Phi_i(k \leftarrow h)$ is invariant with time, and in practice that the geometry does not change with time during the irradiation process. This is quite a strong hypothesis that ought to be ascertained.

This leads to the well-known 'simplified' MIRD equation:

$$\bar{D}_{(k\leftarrow h)} = \tilde{A}_h \times S_{(k\leftarrow h)} \tag{8}$$

As the irradiation of a given target *k* can result from the contribution of several sources, the practical simplified MIRD equation is given by:

$$\bar{D}_k = \sum_h \tilde{A}_h \times S_{(k\leftarrow h)} \tag{9}$$

Note that $S_{(k\leftarrow h)}$ is specific to the radionuclide and does not depend on radiopharmaceutical pharmacokinetics, that is, one set of $S_{(k\leftarrow h)}$ computed for ^{99m}Tc can be used for all radiopharmaceuticals labelled with ^{99m}Tc. This formalism was mostly developed to give an account of the radiation delivered to patients or workers in a situation of deliberate nuclear medicine or accidental occupational internal administration of radionuclides.

The development of computing dosimetric models (reference humans) and the calculation of $S_{(k\leftarrow h)}$ for several radionuclides allowed the assessment of the dosimetric properties of most radiopharmaceuticals available on the market.

However, the formalism is generic and can be applied to a range of situations, for example, for cellular or small animal dosimetry.

6.4. DOSIMETRY OF PENETRATING VS NON-PENETRATING RADIATION

An important concept in radiopharmaceutical dosimetry relates radiation range and the dimensions of a volume of interest. Radiations are called non-penetrating when all emitted energy in a given volume is absorbed in the volume itself. For most organs and tissues (at a clinical scale) this happens with alpha and beta particles and Auger electrons but not with gamma rays. As a consequence, the absorbed fraction for non-penetrating radiation (ϕ_{np}) is 1 when the source and target are confounded and 0 when the source and targets are different:

$\phi_{\mathrm{np}}(k \leftarrow h) = 1$ when $h = k$ (or just: $\phi_{\mathrm{np}}(h \leftarrow h) = 1$)

$\phi_{\mathrm{np}}(k \leftarrow h) = 0$ when $h \neq k$

Conversely, penetrating radiation such as gamma and X rays deposit their energy inside and outside the source region, and therefore the absorbed fraction for penetrating radiation (ϕ_{p}) is:

$\phi_{\mathrm{p}}(k \leftarrow h) < 1$ (and this is also true if $h = k$)

Note that, in fact, radiations can never be completely non-penetrating due to radiative emissions, for example bremsstrahlung. Yet, the definition of penetrating/non-penetrating radiation is conceptually interesting, especially in the context of targeted radionuclide therapy. In targeted radionuclide therapy, the objective is to use non-penetrating radiation as much as possible, in order to irradiate only the tumours where radioactivity is expected to concentrate, while sparing surrounding tissues.

Also note that penetrating/non-penetrating behaviour is a characteristic not of the radiation, but of the relationship between radiation range and the geometry considered. For example, the β^- particles of ^{131}I, with a range of ~1 mm, can be considered to be non-penetrating when the liver is taken as a whole or even for a SPECT voxel (4–5 mm side) in a liver scintigraphy of a ^{131}I labelled compound.

However, the same β^- particles emitted by ^{131}I are definitely penetrating if the volume of interest is a hepatic cell with a diameter of ~10 µm.

This is illustrated in Fig. 13, which represents the variation of the absorbed fraction for water spheres of variable radius homogeneously filled with radionuclides [167].

(a) For very small spheres, regardless of the radionuclide, the energy emitted in the sphere is mostly absorbed outside the sphere, and therefore the absorbed fraction is low (close to 0).
(b) When the sphere radius increases, for the same radionuclide, more and more energy emitted in the sphere is absorbed in the sphere itself, and the absorbed fraction increases.
(c) For even larger spheres, most energy emitted in the sphere is absorbed in the sphere: the absorbed fraction tends towards 1.
(d) When the absorbed fraction tends towards 1, this means that radiation emitted by the radionuclide can be considered to be non-penetrating for the geometry considered.
(e) Last, this behaviour is observed first for low energy emitters, and therefore from the graph one can sort the radionuclides by increasing energy: ^{177}Lu, ^{153}Sm, ^{186}Re and ^{188}Re.

6.5. CELLULAR DOSIMETRY

Cellular dosimetry provides an order of magnitude of the absorbed dose received in cell experiments and may help in comparing the outcomes of experiments using different vectors or radionuclides. The relevance of cellular dosimetry in the context of therapy is also to derive an objective index (the absorbed dose) that can eventually be connected to the observed effect (cell death). Establishing the absorbed dose per event relationship is essential in radiobiology and contributes to the development of a better understanding of the effect of radiation.

Cellular dosimetry is essentially model based dosimetry. The fact is that it is just not possible to follow the fate of a radiopharmaceutical and its effect on individual cells. As cumulated activities are usually obtained as an average over thousands of cells or more, S values are computed for cell models that can be more or less refined.

The easiest way to define a cell is to consider a perfect sphere made of water, surrounded by an infinite water medium. The cell nucleus can be represented by a concentric sphere inside the cell.

The MIRD committee has published a book that gives S values for monoenergetic α particles (3 MeV to 10 MeV) and β^- particles (1 keV to 1 MeV), and for most alpha and beta emitters considered in TRT [168]. Source radii vary from 2 to 10 μm, whereas target radii vary from 1 to 10 μm. Source regions can be selected as cell, cell surface, nucleus or cytoplasm, under the hypothesis of homogeneous distribution of the radioactivity in the source regions. Target regions include cell, nucleus or cytoplasm, and the mean absorbed dose per decay is computed for each target region (Fig. 14).

The absorbed fraction for particulate radiation can be obtained via:

$$\phi_i(k \leftarrow h) = \int_0^\infty \psi_{k\leftarrow h}(x) \frac{1}{E_i} \left. \frac{\mathrm{d}E}{\mathrm{d}X} \right|_{X(E_i)-x} \mathrm{d}x \tag{10}$$

where

$X(E_i)$ is the range of that particle;

$\psi_{k\leftarrow h}(x)$ is a geometric reduction factor;

$\left. \frac{\mathrm{d}E}{\mathrm{d}X} \right|_{X(E_i)-x}$ is the stopping power evaluated at the residual range;

and $X(E_i)-x$ is the residual range of the particle that has passed a distance x in the medium.

The geometric reduction factor is calculated for different source/target configurations and can be seen as the probability that a particle emitted in the source hits the target. For example, when the source is the cell surface (CS) and the target is the cell nucleus (N):

$$\psi_{\mathrm{N}\leftarrow\mathrm{CS}}(x) = \begin{cases} 0 & \text{when } 0 \le x \le R_\mathrm{C} - R_\mathrm{N} \\ \dfrac{2xR_\mathrm{C} - R_\mathrm{C}^2 - x^2 + R_\mathrm{N}^2}{4xR_\mathrm{C}} & \text{when } R_\mathrm{C} - R_\mathrm{N} \le x \le R_\mathrm{C} + R_\mathrm{N} \\ 0 & \text{when } x \ge R_\mathrm{C} + R_\mathrm{N} \end{cases} \tag{11}$$

The evaluation of the stopping power is made according to the energy–range relationship proposed by Cole for electrons (at least for electron energies above 0.4 keV — i.e. a range of 19 nm) [169]. For alpha particles, stopping power values are interpolated from ICRU Report No. 49 [170].

The S values presented in the MIRD book [168] have been used extensively to compare the energy delivered by different radionuclides in cellular geometry. The book also presents an interesting discussion on the assumptions and limitations of the approach. For example, the dependence of the S value on

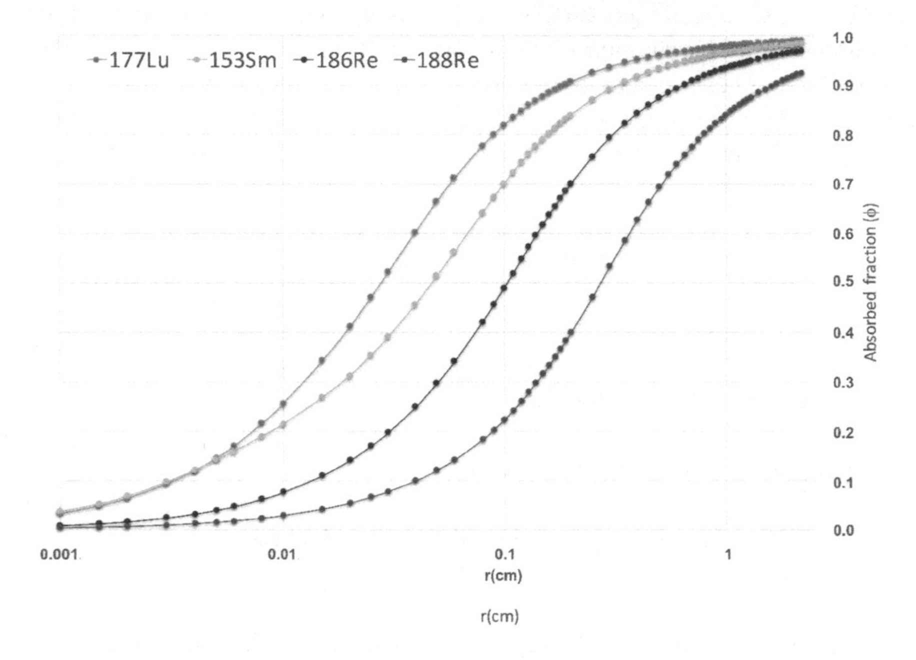

FIG. 13. Absorbed fractions for spheres of variable radii r(cm) homogeneously filled with four radionuclides (^{177}Lu, ^{153}Sm, ^{186}Re, ^{188}Re). Data adapted from Ref. [167] (courtesy of M. Bardiès, Institut de Recherche en Cancérologie de Montpellier).

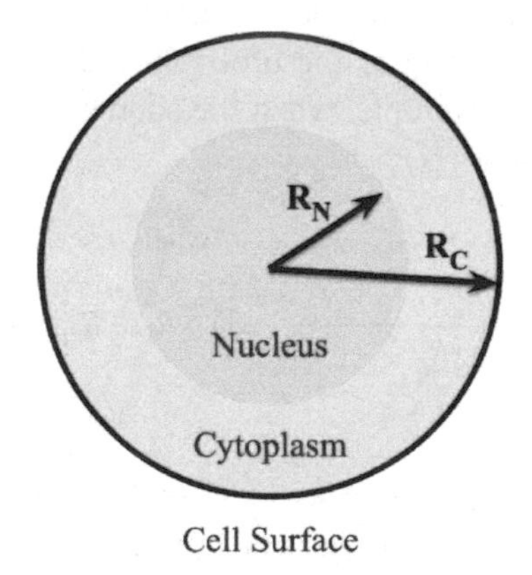

FIG. 14. Cell geometry is represented as a sphere. RC and RN are the cell and nucleus radii, respectively (courtesy of M. Bardiès, Institut de Recherche en Cancérologie de Montpellier).

cell geometry (ellipsoid vs perfect sphere) is studied. As can be expected and for geometrical (ballistic) reasons, the impact of cell geometry changes is more pronounced when sources and targets are different (e.g. *S* values for nucleus from cytoplasm, of cell from cell surface). However, the impact is less pronounced when sources and targets are distant.

An important point to consider when using *S* values is that when radioactive decay leads to unstable decays, the whole decay chain ought to be considered for each radionuclide, with the branching ratio, and taking the dynamics of radioactive decay into account.

More recently, an online tool (MIRDcell[2]) was made available that allows the user to consider not only isolated cells but also cell clusters arranged in different structures [171]. The calculation of the absorbed dose is made according to the reference cited above, and the generalization from single isolated cell to cell clusters is derived from another study from Goddu et al. [172]. Variable cell labelling patterns can be modelled for cell colonies that develop in plane or in

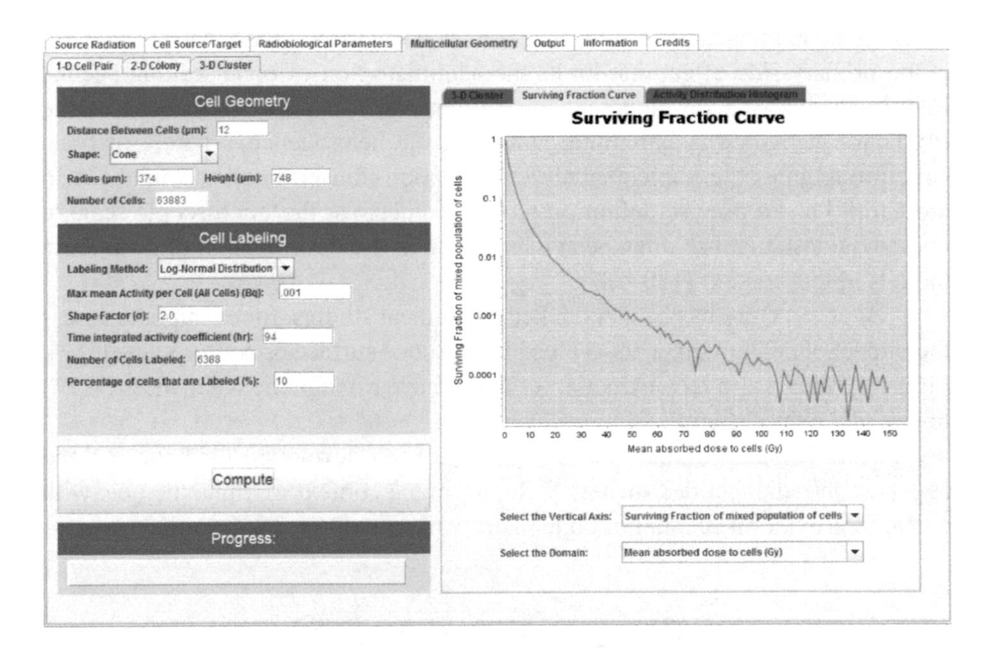

FIG. 15. Screenshot of MIRDcell: plot of survival fraction of cells in multicellular geometry (3D cluster) (reproduced from Ref. [171] with permission).

[2] See http://mirdcell.njms.rutgers.edu

3D. In addition, MIRDcell enables the user to model the surviving fraction of cells according to the linear quadratic model (Fig. 15).

Obviously, these tools are mostly designed to provide an order of magnitude of the irradiation delivered. Depending on the objective — for example, for radiobiological experiments — more refined dosimetric approaches may have to be implemented [173–175]. The development of microdosimetric approaches that take the stochastic nature of energy deposition into account is also possible when relevant [176–178]. The fact is that the possibilities for modelling radiation transport and energy deposition even at the microscopic scale have improved markedly in recent decades, and can contribute to a better understanding of radiation effects at the cellular scale — as long as geometry and radiopharmaceutical pharmacokinetics can be assessed with a sufficient degree of accuracy [179].

6.6. TISSUE AND SMALL SCALE DOSIMETRY

As bone metastases are located near bone marrow, bone marrow toxicity is the primary side effect that limits the administration of bone seeking agents. In this context, the development of tissue models to give an account of the irradiation delivered is important. However, the heterogeneous nature of bone structure requires the implementation of radiation transport codes (Monte Carlo modelling) and a precise definition of the geometry at the microscopic scale, a non-trivial task that has often been a limitation in the development of dosimetric models in that context [180–182].

Bone structure has been modelled for clinical studies, for example to assess the impact of radiopharmaceutical location on bone surface or bone volume [183]. Figure 16 shows the reconstruction of a 3D nuclear magnetic resonance (NMR) image of a cubical sample of trabecular bone.

More refined microscopic models have been developed for bone seeking agents with radionuclides such as ^{186}Re, or, in the context of alpha therapy, with ^{223}Ra, to provide an account of bone marrow toxicity [184, 185].

6.7. APPLICATION TO SMALL ANIMAL STUDIES

Another aspect of radiopharmaceutical development that can benefit from dosimetric studies is small animal studies. Collecting dosimetric data from small animals is required before considering any clinical studies. Most of the time, pre-clinical studies are used to assess animal pharmacokinetics, extrapolate to human pharmacokinetics (allometric scaling) and assess the absorbed doses that could be delivered to human dosimetric models [186, 187]. Codes such as OLINDA [188, 189] or IDAC [190] are used for that purpose. In targeted radionuclide therapy, pre-clinical (small animal) studies can also be used to study the effect of the therapy (efficacy or toxicity). The situation is often complicated, depending on the experimental animal model used [191]. Xenografts of human origin can be grown in immunosuppressed hosts, such as genetically manipulated athymic mice. The efficacy of the radiopharmaceutical can be studied on human tumours grown in mice, making the results easier to transpose to humans. However, the toxicity estimated will be that for the animal and hence less informative regarding what could potentially happen to humans. Hence, experimental targeted radiotherapy performed on small animals is a domain where dosimetry can be implemented, as in clinics, to study the relationship between irradiation delivered and observed effects.

Small animal dosimetric models have been proposed and, as in clinics, mathematical models preceded voxel based models [192]. Figure 17 depicts the mouse phantom MOBY [193]. There are some warnings associated with the use of murine dosimetric models. These relate to the fact that the size of most organs is of the same order of magnitude as the radiation range. As a result, small differences in the geometry may lead to important absorbed dose gradients [194, 195]. This can raise questions about the relevance of model based dosimetry if the specific animal geometry deviates from the selected model.

Another aspect not explicitly treated here is how radiopharmaceutical pharmacokinetics can be assessed. The usual method is the 'cut and count' approach, in which tissue samples are collected after animal sacrifice and weighed and counted to derive the activity concentration at various time points. Mouse

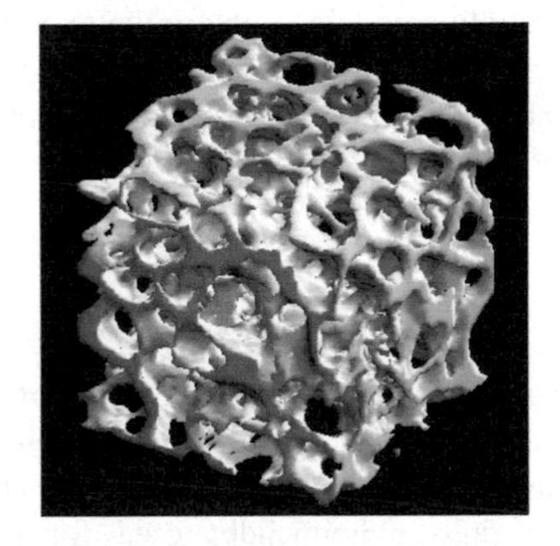

FIG. 16. Reconstruction of a 3D NMR image of a cubical sample of trabecular bone. The sample size is 5.6 × 5.6 × 5.6 mm^3 and cubic voxel sampling is 88 mm^3 (reproduced from Ref. [182] with permission).

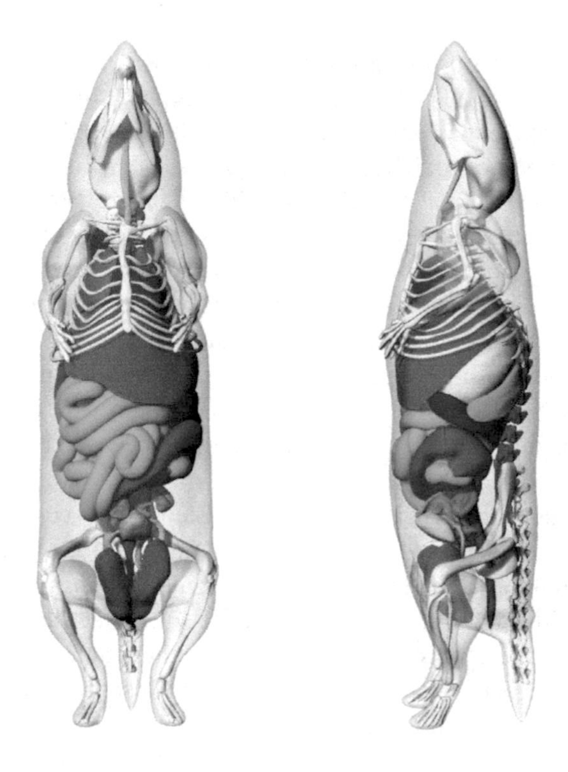

FIG. 17. Lateral (right) and anterior (left) views of the digital mouse phantom MOBY (courtesy of P. Segars, Duke University).

specific pharmacokinetics can be obtained from sequential quantitative imaging of the same animal. However, activity quantification in pre-clinical imaging devices needs to be carefully ascertained before the experiments are conducted. When carried out carefully, pre-clinical dosimetry results are very useful in improving our basic knowledge of radiation effects, as well as in estimating the relative effectiveness of TRT using different radiopharmaceuticals.

6.8. CONCLUSION

Pre-clinical dosimetry is an essential part of radiopharmaceutical development. It is essential to study the radiobiological outcome of radiopharmaceutical dose delivery. It is also used to compare the effect of radiation emitted by various radionuclides used for targeted therapy with radiation dose delivered by external beams.

The possibilities for deriving the absorbed dose have improved markedly within recent decades. However, a lack of expertise in the field of pre-clinical dosimetry limits its application to pre-clinical experiments in many research groups.

Most pre-clinical dosimetry is derived from dosimetric models (of cells, tissues, small animals), but it is also possible under some circumstances to refine the dosimetric approach, depending on experimental requirements. In fact, the biological experiment and the expected outcome always condition the dosimetric approach that will need to be implemented in each case. This means that multidisciplinary collaborations are essential in radiopharmaceutical development.

7. BONE PAIN PALLIATION RADIOPHARMACEUTICALS

7.1. CURRENT STATUS OF BONE PAIN PALLIATION AGENTS

Certain cancers, such as prostate, breast and lung, commonly spread to other parts of the body as the disease advances. Improvement in quality of life is the main objective of palliative care for terminally ill patients with bone metastasis [196–198]. The severe pain due to the wide spread of disease in the skeleton is managed using different drugs; bone seeking radiopharmaceuticals play an important role because of their minimal side effects [43, 199, 200]. In spite of the fact that a large number of radiolabelled products targeting bone tissues have been developed and tested in laboratory animals and also in suitable animal models, only a handful of these agents have found regular use in palliative care for cancer patients [201]. This section describes the bone seeking radiopharmaceuticals whose systematic use has been documented or proposed for bone pain palliation.

7.2. BONE PAIN PALLIATION AGENTS WITH WELL ESTABLISHED CLINICAL USE

7.2.1. ^{32}P

Sodium orthophosphate containing radioactive ^{32}P ($E_{\beta(max)}$ = 1.711 MeV, no γ emission, $T_{1/2}$ = 14.26 days) was the first radiopharmaceutical used for metastatic bone pain palliation [202]. The radiopharmaceutical was reported to be administered in patients either intravenously or orally. The administered activity generally varies based on the route of administration of the radiopharmaceutical. Usually, an activity of 185–370 MBq (5–10 mCi) is injected intravenously.

Phosphorus-32 is also given orally and a higher activity of 370–444 MBq (10–12 mCi) is usually administered [203]. The accumulation of ^{32}P in the bone is due to the affinity of phosphate groups for the calcium present in the bone to form the particles of HA, which is the major constituent of the bone matrix. The onset of pain relief starts ~14 days after administration of the radiopharmaceutical. From patient to patient, it could vary from 2 days to 4 weeks and the patients could have pain free survival for almost 2 to 4 months. Pain relief has been reported by 59–93% of patients suffering from prostate cancer, with an overall response rate of 77%, while 52–94% of patients suffering from breast cancer have reported pain relief, with an overall response rate of 84% [204].

The formulation of the agent is quite simple, as it is administered in the form of sodium orthophosphate, which eliminates the requirement for a sophisticated radiopharmaceutical formulation facility. Phosphorus-32 has a reasonably long half-live that permits shipment of the product to locations that are far from the production site. Cost effective production and a comparatively long half-life make ^{32}P quite economical and readily available to a wider patient population [70, 97].

As an endogenous elemental radionuclide of the bone inorganic matrix, ^{32}P was used extensively until the 1980s, but its myelotoxicity due to the emission of high energy β^- particles was a disadvantage [199, 204–206]. Phosphorus-32 is a pure β^- particle emitter, and pharmacokinetic evaluation and dosimetric assessment in patients during therapy are therefore not possible.

7.2.2. $^{89}SrCl_2$ (metastron)

Strontium-89 ($E_{\beta(max)}$ = 1.49 MeV, no γ emission, $T_{1/2}$ = 50.5 days), in the form of strontium chloride (also known as metastron), is one of the best established agents used for metastatic bone pain palliation. The agent is administered intravenously, either in an activity range that varies from 1.5 to 2.2 MBq/kg (40 to 60 μCi/kg) of the patient's body weight, or as a cumulative activity of 148 MBq (4 mCi) [207]. Strontium and calcium are group 2 elements in the periodic table; hence, ^{89}Sr behaves similarly to calcium in the biological system. Thus, ^{89}Sr is taken up by the bone matrix in proportion to the osteoblastic activity [67, 208]. The onset of pain relief is observed 2–4 weeks after the administration of $^{89}SrCl_2$, with a treatment efficiency of 60–84%. Thanks to its reasonably long half-life, patients can feel pain relief for 4–15 months with $^{89}SrCl_2$ treatment [204, 209]. In pre-clinical and clinical studies skeletal tissues exhibit nearly 20% retention of injected activity even 3 months after administration, overcoming the limitation of the very short biological half-life of the agent [210].

The radiopharmaceutical is administered as an inorganic preparation of $^{89}SrCl_2$ and thus its formulation is relatively simple. The long half-life of ^{89}Sr

provides a logistical advantage, making it possible to distribute the agent to any part of the globe without much concern about the loss of activity due to radioactive decay. Additionally, the requirement of only ~148 MBq (~4 mCi) of activity per patient reduces the cumulative activity requirement, making $^{89}SrCl_2$ an attractive choice for bone pain palliation applications. For this reason, $^{89}SrCl_2$ is considered to be a good choice for patients with moderate pain with somewhat longer life expectancy.

Although $^{89}SrCl_2$ has proven to be an efficacious agent for bone pain palliation, the limited availability and high cost of the agent are serious impediments for its widespread use. The difficulty of large scale production of ^{89}Sr is the primary reason for the high cost of $^{89}SrCl_2$ [70].

7.2.3. ^{153}Sm–EDTMP (quadramet)

Samarium-153 labelled ethylenediaminetetramethylene phosphonic acid (^{153}Sm–EDTMP), commercially known as quadramet, is the most widely used radiopharmaceutical for metastatic bone pain palliation; ^{153}Sm–EDTMP is also known as lexidronam. The agent is administered intravenously with an activity of 37 MBq/kg (1 mCi/kg) of the patients' body weight [204, 211–213]. The agent is accumulated in the skeleton owing to the affinity of the phosphonate groups, present in EDTMP, for calcium in the bone matrix. The accumulation of the agent is usually proportionate to the osteoblastic activity. Pain reduction is usually experienced within 2 to 7 days after administration. The agent is reported to provide a pain free survival period varying from 4 weeks to 35 weeks, with a mean pain free period of ~8 weeks. ^{153}Sm–EDTMP is reported to be effective in 62–74% of cancer patients [212, 214, 215]. Figure 18 depicts scintigraphic planar images taken in a patient administered ^{153}Sm–EDTMP (left) and shows that the radiopharmaceutical is taken up by the metastatic lesions identified with ^{99m}Tc–MDP (right) (MDP acid).

One of the major advantages of using ^{153}Sm–EDTMP is the fact that it is derived from ^{153}Sm, which can be produced by employing the simple (n,γ) route using medium flux research reactors. The radionuclide is usually produced with high radionuclidic purity and adequate specific activity using an enriched (in ^{152}Sm) Sm_2O_3 target owing to the large thermal neutron capture cross-section of ^{152}Sm (σ = 208 b and resonance integral 2978 b) [78]. It has been documented that ^{153}Sm can also be produced with adequate specific activity and sufficient radionuclidic purity for bone pain palliation application using natural Sm_2O_3 as a target material [100].

The attractive nuclear decay characteristics of ^{153}Sm, namely moderate energy β^- emission, ensure lower bone marrow toxicity, and the availability of accompanying suitable energy γ photons helps in performing simultaneous scintigraphic evaluation and dosimetry studies. Moreover, a simple formulation for the radiopharmaceutical, either by using a freeze-dried EDTMP kit or by following in situ wet chemistry protocols at the hospital radiopharmacy, makes ^{153}Sm–EDTMP an attractive agent for large scale commercial deployment [201].

One of the problems associated with ^{153}Sm based agents is the short half-life of ^{153}Sm, which requires the handling of large quantities during production in order to supply the radionuclide or radiolabelled agents to locations that are distant from the radionuclide production or radiopharmaceutical formulation site. Moreover, the amount of activity required per patient dose is also comparatively large compared to other agents, such as $^{89}SrCl_2$.

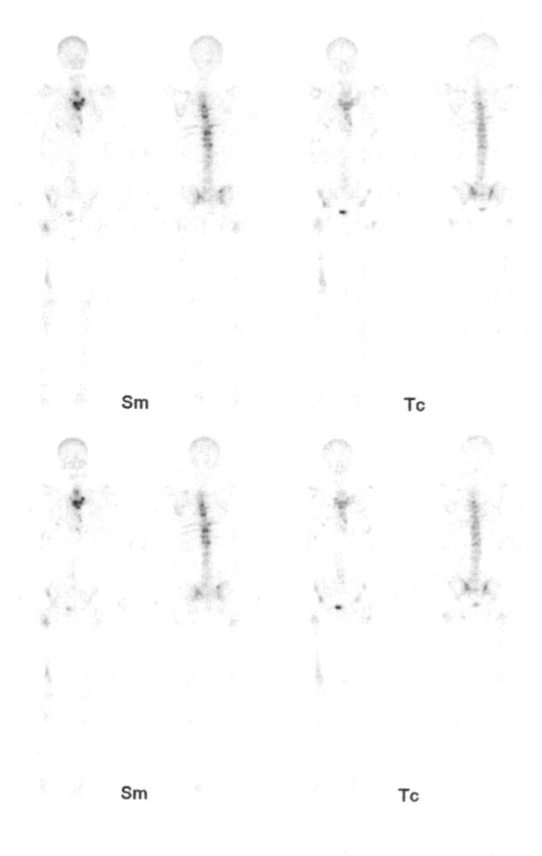

FIG. 18. Scintigraphic images of a patient administered ^{153}Sm–EDTMP (left) and ^{99m}Tc–MDP (right) (courtesy of A. Sasikumar, KIMS Hospital DDNMRC).

7.2.4. ^{188}Re HEDP

Both of the β^- emitting radionuclides of rhenium, namely ^{186}Re and ^{188}Re, are used for the preparation of bone pain palliation radiopharmaceuticals [216]. However, more clinical studies have been reported with ^{188}Re thanks to its availability from a ^{188}W/^{188}Re generator. Rhenium-188 labelled HEDP has been used for bone pain palliation. Rhenium-188 HEDP is administered through intravenous injection with an activity that varies from 2.96 to 4.44 GBq (80–120 mCi) [217–219]. Repeated doses have been shown to be more effective for palliation and patient survival [220]. The radiopharmaceutical accumulates in bone matrix by binding to calcium through the phosphate group present in HEDP [221]. It is reported that the presence of the hydroxyl group in HEDP enhances the accumulation of the radiolabelled agent in the skeleton [198]. The agent is reported to be effective in 60–95% of patients [217, 219, 222].

The availability of ^{188}Re from a ^{188}W/^{188}Re radionuclide generator, which can be housed in a central or hospital radiopharmacy, allows ^{188}Re HEDP to be formulated at any point of time using freeze-dried HEDP kits [223–226]. This also provides hospitals with a means to treat patients when the supply of other reactor produced radioisotopes (^{89}Sr, ^{153}Sm, ^{177}Lu, etc.) is disrupted. The presence of 155 keV gamma photons does not substantially enhance the irradiation delivered to patients, but enables simultaneous pharmacokinetic evaluation and dosimetry studies.

However, the widespread availability of the agent may suffer significantly from the fact that very high flux reactors are essential to produce the quantities of ^{188}W required for commercial exploitation. Tungsten-188, the parent radionuclide required for the preparation of the ^{188}W/^{188}Re radionuclide generator, is produced by a double neutron capture reaction using enriched ^{186}W as the target. Extended irradiation of the target in nuclear reactors with thermal neutron fluxes in the range of $\sim 10^{15}$ $n \cdot cm^{-2} \cdot s^{-1}$ is required [87]. The limited availability of such reactors globally makes the ^{188}W/^{188}Re generator, and consequently ^{188}Re, quite expensive, and this may be a serious impediment to the widespread availability of this otherwise useful agent [70]. Moreover, the significantly higher energy (2.12 MeV) β^- emission is a cause of serious concern, as it may lead to significant myelotoxicity and bone marrow suppression.

7.2.5. $^{223}RaCl_2$ (alpharadin) (xofigo)

Radium-223 chloride, also known as xofigo, is an α particle emitting bone pain palliation agent, the clinical use of which was approved by the Food and Drug Administration (FDA) of the USA in 2013. As radium is in group 2 of the periodic table, it has similar properties to calcium and strontium and hence is accumulated in the skeleton [127, 227, 228]. The radiopharmaceutical is administered through intravenous injection, either as a single administration of 200–250 kBq/kg (5.4–6.8 μCi/kg) of a patient's body weight or in multiple injections involving five to six administrations of ~55 kBq/kg (1.49 μCi/kg) of the patient's body weight [44, 229, 230]. The product is reported to not only control bone pain, but also be successful in providing a few additional months of survival to the terminally ill cancer patients [230, 231].

Figure 19 shows SPECT images taken in a patient administered $^{223}RaCl_2$ (left) and the corresponding ^{99m}Tc–MDP images (right). The chain of ^{223}Ra provides two gamma rays, E_γ = 269 keV (13.6%) and 154 keV (6.04%), which are useful for imaging. However, the amount of the radiopharmaceutical injected into the patient is very low — a few hundred kBq — and the quality of the images will not be good despite long imaging times.

One of the basic advantages of using ^{223}Ra for bone pain palliation is that it is used in the simple inorganic form of $^{223}RaCl_2$ and thus the formulation of the

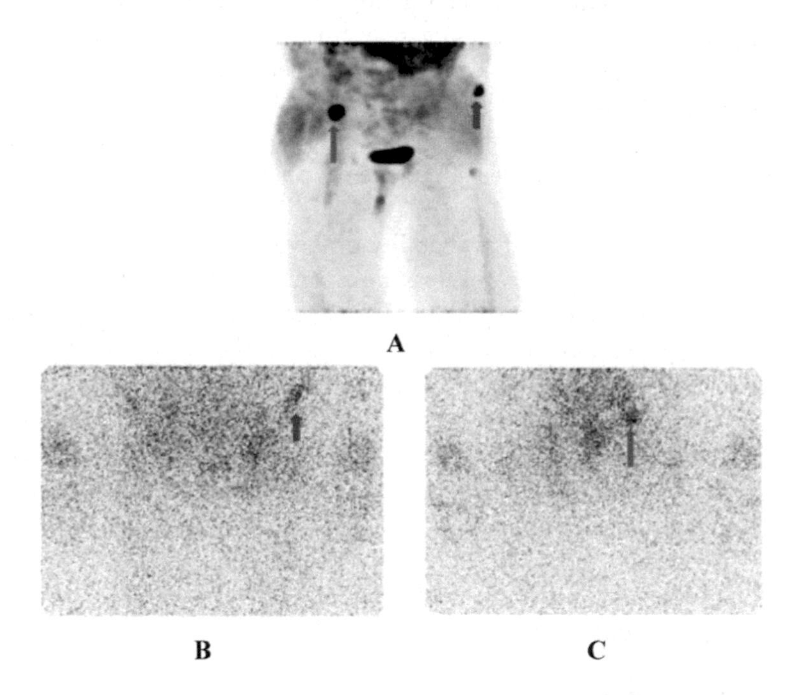

FIG. 19. Diagnostic imaging of a patient treated with xofigo (bottom) and pre-therapy F-18-fluorocholine-PET (up). PET images showed strong uptake in the right iliac crest (blue arrow) and left anterior ileal wing (red arrow) (a). The same lesions were imaged at 7 days after xofigo using a gamma camera: the left anterior ileal wing (red arrow) lesion is rendered better in the AP view (b), whereas the lesion in the right iliac crest (blue arrow) is rendered better in the AP view (c) (reproduced from Ref. [232] with permission).

agent is simple and convenient. The long half-life ($T_{1/2}$ = 11.4 days) is useful, as the radiopharmaceutical can be prepared and shipped globally.

Radium-223 is separated from an equilibrium mixture of the ^{227}Ac–^{223}Ra parent–decay system. The parent radionuclide ^{227}Ac is produced by prolonged neutron irradiation of natural ^{226}Ra in a nuclear reactor [133]. Radium-223 decays via a chain of short lived decay radionuclides, producing four α particles, to stable ^{207}Pb, with total decay energy of 28 MeV [44, 233]. As the range of α particles is very short, the irradiation is delivered to the metastatic bone where the radionuclide is taken up initially with negligible irradiation to the underlying bone marrow.

The longer half-life of ^{223}Ra provides sufficient time for preparation of the agent, quality control studies and distribution to distant locations. The accompanying low energy gamma photons of ^{223}Ra are suitable for performing scintigraphic imaging, albeit with poor statistics, and possibly dosimetric calculations. However, as the amount of radioactivity injected is small, it is often difficult to obtain good images. The low energy gamma photons do not induce much external irradiation; the treatment can thus be performed on an outpatient basis [234].

In spite of the exciting clinical results reported to date, widespread use of this radiopharmaceutical appears distant owing to its prohibitively high cost. Moreover, the clinical data for use of the agent are also limited and predominantly based on a single large multi-centre clinical trial [231].

7.3. BONE PAIN PALLIATION AGENTS THAT ARE CLINICALLY TESTED

7.3.1. ^{186}Re HEDP (etidronate)

Rhenium-186 labelled HEDP is one of the first and so far most widely used radiometal labelled bisphosphonates for palliative radiotherapy of painful osseous metastases [235, 236]. ^{186}Re HEDP has been approved in some European countries for regular clinical use as a pain palliative radiopharmaceutical for skeletal metastases. The agent is administered intravenously with an activity ranging from 1.48 to 3.33 GBq (40 to 90 mCi). The bio-accumulation pathway of the agent is similar to that of ^{188}Re HEDP. The limited number of clinical studies performed with the agent to date showed it to be effective in 77–90% of cancer patients [237, 238].

In spite of the extensive clinical use of $^{186/188}Re$ HEDP, these agents have some inherent drawbacks. The most important of these is the fact that Re–bisphosphonate complexes have a strong tendency to oxidize to $[ReO_4]^-$ in vivo and this in turn reduces skeletal uptake and increases the accumulation of activity in non-target organs such as the thyroid. Moreover, Re–bisphosphonate complexes do not have a well-defined chemical structure.

7.3.2. ^{117m}Sn DTPA

The use of very low energy electrons for bone pain palliation in the form of ^{117m}Sn labelled DTPA was first proposed and clinically demonstrated by Srivastava et al. [239–241]. The agent is administered intravenously and reported to provide substantial pain relief to patients within 1 week of administration when injected, with an activity of >444 MBq (12 mCi) [242]. The mechanism of uptake of this tin complex is not well understood. It is proposed that bio-accumulation of ^{117m}Sn–DTPA in the skeleton occurs either due to the precipitation of stannous oxide on the bone surfaces or by a hydrolysis reaction with HA particles, one of the primary constituents of the bone matrix [204]. Clinical studies performed by Srivastava et al. showed that the agent is effective in controlling bone pain in 60–83% of patients [241, 242].

The palliative effects exhibited by the agent are due to the irradiation of the bone matrix by the low energy conversion electrons emitted by ^{117m}Sn (conversion electrons of 127 and 129 keV, E_γ = 159 keV (86%), $T_{1/2}$ = 13.6 days). The conversion electrons are of very low energy and reduce the irradiation to the bone marrow and consequently induce low myelotoxicity, which is considered to be the major advantage of using this agent [243]. Further, the presence of suitable energy mono-energetic gamma photons of 159 keV (86%) enables simultaneous scintigraphic imaging and pharmacokinetic studies. The relatively long half-life of ^{117m}Sn means that there are also advantages of production and supply logistics for this agent without much loss of activity through radioactive decay.

One of the major disadvantages of using ^{117m}Sn–DTPA for bone pain palliation arises from the production of ^{117m}Sn. It is difficult to produce ^{117m}Sn in the majority of nuclear reactors available worldwide. The radionuclide is produced either by thermal neutron bombardment of ^{116}Sn by the nuclear reaction ^{116}Sn(n,γ)^{117m}Sn or by inelastic neutron scattering following the ^{117}Sn(n,n′)^{117m}Sn nuclear reaction. Both the nuclear reactions have poor cross-sections and thus the use of very high flux reactors (~10^{15} n·cm^{-2}·s^{-1}) and highly enriched targets (the natural abundances of ^{116}Sn and ^{117}Sn are 14.7% and 7.68%, respectively) are mandatory for its production in adequate quantities. Although ^{117m}Sn can also be produced in NCA form in particle accelerators using a ^{3}He particle induced reaction on ^{116}Cd or ^{115}In target, these nuclear reactions also suffer from low yields [244, 245]. This makes the production of ^{117m}Sn, and thus ^{117m}Sn–DTPA, quite expensive, and this is a serious drawback for large scale utilization of this otherwise useful agent. Hence, only limited clinical experience with this agent has been reported to date.

7.3.3. ^{177}Lu–EDTMP

^{177}Lu–EDTMP is a radiopharmaceutical developed for bone pain palliation that has undergone extensive pharmacokinetic evaluation in different animal species as well as clinical use thanks to the initiative of the IAEA through its CRP [246]. The radiopharmaceutical is formulated using a freeze-dried kit containing 35 mg of EDTMP, 5.72 g of Ca^{2+} ions and 14.1 mg of NaOH. The kit is reconstituted with 1 mL of normal saline and ~3.7 GBq (100 mCi) of $^{177}LuCl_3$ is added to it before incubation at room temperature for 5–10 min. Detailed in vivo evaluation studies were conducted using this formulation in mice, rats, rabbits and dogs. ^{177}Lu–EDTMP accumulated almost exclusively in the skeletal system (peak uptake ca. 41% of the injected activity in bone for 1–3 h with a terminal elimination half-life of over 80 days). That part of the radiopharmaceutical that was not taken up by the bone was excreted through the urinary system. The uptake in the bone remained constant without much redistribution or excretion.

All species (mouse, rat, rabbit, dog) studied showed uptake of the radiopharmaceuticals in the bone. The kidneys were free of any radioactivity after one day of administration of the radiopharmaceutical. A moderate decrease of platelet concentration (160 g/L) was observed at one week post-administration in dogs with the highest administered activity of 37 MB/kg. However, no toxicity was seen. The effective half-life of ^{177}Lu–EDTMP was long, indicating that the complexation of ^{177}Lu with EDTMP did not affect the biological behaviour of the ligand, EDTMP, as a bone seeker. Species specific biodistribution behaviour was observed. Detailed toxicity studies in dogs did not indicate toxicity of the radiotracer or any adverse biological effects.

Chakraborty et al. reported promising pre-clinical results for ^{177}Lu–EDTMP in canine patients with primary as well as metastatic bone lesions. ^{177}Lu–EDTMP was used at a ~44.4 MBq (1.2 mCi) per kg body weight dose for these studies [247]. The administered activity of the radiopharmaceutical showed very good therapeutic efficacy. Two dogs suffering from primary bone cancer showed palliative effect as well as stable disease condition after treatment with ^{177}Lu–EDTMP. One of the dogs suffering from skeletal metastases originating from a primary tumour elsewhere survived for nearly four years after treatment with one administration of ^{177}Lu–EDTMP and died from another cause.

The clinical trials for this radiopharmaceutical were conducted in nuclear medicine departments in Member States under the guidance of the IAEA. Tracer activities of 173–207 MBq (4.6–5.6 mCi) of ^{177}Lu–EDTMP were administered to patients suffering from metastatic prostate cancer in the phase −0 study; the activity accumulation rate was studied for the whole skeletal system, other organs, and blood and urine to assess in vivo kinetics [7]. The study showed faster biphasic clearance from blood, prolonged retention in bone tissue and non-accumulated activity being excreted through the kidneys. The average accumulated dose was 0.16 ± 0.04 mGy/MBq for total body and 0.80 ± 0.15 mGy/MBq for red marrow, as estimated using OLINDA/EXM 1.0 software. A maximum tolerated dose of 2000–3250 MBq (54–88 mCi) for ^{177}Lu–EDTMP was inferred from these studies.

For the activity escalation study, 21 patients with metastatic prostate cancer were given ^{177}Lu–EDTMP in the range of 692–5550 MBq (19–150 mCi). Blood cell counts for platelet and leukocytes and haemoglobin levels were monitored for over 12 weeks as toxicity parameters. When doses higher than the maximum tolerated dose (3.26 GBq (88 mCi)) were administered to patients, toxicity grades 3–4 were observed; otherwise, one third of patients showed very good tolerance for normal doses. An efficiency of 86% was observed for pain relief, with an average duration of >7 weeks. The whole study demonstrated the suitability of ^{177}Lu–EDTMP as a safe and effective bone pain palliative agent.

Based on the encouraging dosimetry and toxicity data for ^{177}Lu–EDTMP, a phase II clinical trial was performed in several countries [8, 9, 248]. The radiopharmaceutical was administered intravenously as a single administration of 1.3–3.7 GBq (35–100 mCi) and early onset of pain relief was observed within one week. The pain free survival period was maintained for a period of 2–4 months. The clinical studies performed with the agent to date documented a positive response in 77–95% of patients. A recent clinical study carried out in patients with painful skeletal metastases using either ^{177}Lu–EDTMP or ^{153}Sm–EDTMP on an equidose basis showed that both agents have similar pain response efficacy [249].

The radionuclide decay characteristics of and easy production logistics for ^{177}Lu are the major plus points for ^{177}Lu based theranostic agents. Moderate energy β^- particle emission reduces the possibility of bone marrow suppression and subsequent myelotoxicity. The low energy, low abundance gamma photons are useful for scintigraphic and dosimetric evaluation without causing more irradiation of the patients. The relatively long half-life of ^{177}Lu is advantageous because there is sufficient time for the production and radiochemical processing of ^{177}Lu, as well as the formulation and supply of ^{177}Lu based radiopharmaceuticals across the globe, without much loss of radioactivity.

The very high thermal neutron capture cross-section of ^{176}Lu (σ = 2100 b) ensures the production of ^{177}Lu with high radionuclidic purity and adequate specific activity for the preparation of bone pain palliation agents. Natural Lu_2O_3 targets can be used in medium flux reactors for the production of ^{177}Lu with adequate specific activity for the preparation of bone pain palliation agents. Moreover, the ease with which the patient dose of ^{177}Lu–EDTMP can be formulated, either by following the wet chemistry protocol or by using freeze-dried EDTMP kits, makes it an attractive agent for bone pain palliation [154].

7.3.4. ^{177}Lu–DOTMP

Lutetium-177 labelled DOTMP has been reported as another potential agent for metastatic bone pain palliation [56, 250]. Macrocyclic chelators have advantages over their acyclic counterparts for radiolabelling with lanthanide isotopes such as ^{177}Lu, as these complexes are kinetically and thermodynamically more stable [101, 251]. Compared to the two amine groups in EDTMP, the four amine groups available in these ligands favour better complexation properties. The agent can be prepared at a hospital radiopharmacy, either by a wet chemistry method or by using a freeze-dried DOTMP kit following simple procedures [252, 253]. Preliminary biological studies in small animal models indicated faster clearance of activity from blood and lower retention of activity in the non-target organs for ^{177}Lu–DOTMP compared to ^{177}Lu–EDTMP under identical conditions [254]. Higher animal studies in dogs involving the administration of ^{177}Lu–DOTMP at high activities did not exhibit any clinical toxicity [255]. Clinical studies with the agent reported successful administration of up to 3.7 GBq (100 mCi) of the preparation in cancer patients with preferential localization of the radiotracer in the osteoblastic lesions and almost no uptake in soft tissue or any other major non-target organs [256].

7.3.5. ^{153}Sm–DOTMP

Samarium forms a very stable complex with the DOTMP ligand and hence ^{153}Sm–DOTMP is also one of the potential radiopharmaceuticals for bone pain palliation [252, 254, 255]. Lanthanide complexes with the cyclic ligand, DOTMP, can be prepared with a much lower ligand concentration as compared to the acyclic ligand, EDTMP. Hence, low specific activity ^{153}Sm will suffice for the preparation of ^{153}Sm–DOTMP.

^{153}Sm–DOTMP is under commercial development under the trade name CycloSam. It uses a 3 : 1 ligand to metal ratio, although a 1 : 1 ratio is only required for complex formation in high yields. Quadramet, when manufactured commercially, requires a 300 : 1 ligand to samarium ratio to ensure that samarium remains chelated to EDTMP in vivo. CycloSam only localizes in bone with almost no uptake elsewhere in the body and it clears quickly through the renal system.

7.3.6. ^{170}Tm–EDTMP

Thulium-170 is a medium energy ($E_{\beta max}$ = 968 keV) β^- particle emitting radionuclide and is an interesting choice for developing agents for bone pain palliation. It has a half-life of 128.4 days, which is considerably longer than that of the radionuclides currently used for bone pain palliation; the longest lived of these is ^{89}Sr, with a half-life of 50 days. The medium energy β^- particle is beneficial, as it is expected to spare bone marrow irradiation. The gamma rays can be used for SPECT imaging to study the pharmacokinetics and dosimetry. This radionuclide was originally proposed by Das et al., who prepared ^{170}Tm–EDTMP and studied its pharmacokinetics in Wistar rats [79]. The ^{170}Tm used in these studies was produced by thermal neutron bombardment of a natural Tm_2O_3 target for a period of 60 days at a flux of 6×10^{13} $n \cdot cm^{-2} \cdot s^{-1}$. The specific activity of the product was 6.36 GBq (170 mCi)/mg and the radionuclidic purity was close to 100%.

Thulium-170-labelled EDTMP was prepared by incubating 1 mg of EDTMP, dissolved in bicarbonate buffer, with 111–185 MBq (3–5 mCi) of $^{170}TmCl_3$ at room temperature. The radiolabelled complex was obtained with >99% radiochemical purity and exhibited excellent stability for several days at room temperature. Biological evaluation involving biodistribution studies and scintigraphic imaging, performed in normal Wistar rats, showed significant accumulation of the radiotracer in the skeleton (50–55% of the administered activity) along with prolonged bone retention for the duration of the studies up to 60 days post-administration. Dosimetric evaluation revealed that the agent delivered a dose of 4.3 Gy/MBq to the skeleton, while the corresponding absorbed dose to the red bone marrow was only ~0.5 μGy/MBq. These studies indicated that ^{170}Tm–EDTMP has excellent biological features and favourable dosimetric values, making it a promising agent for metastatic bone pain palliation.

The developers suggested the use of ^{170}Tm–EDTMP as a cost effective alternative to $^{89}SrCl_2$. Unlike with ^{89}Sr, large quantities of ^{170}Tm can even be prepared at medium flux research reactors, while thulium is monoisotopic and thus enrichment is not necessary [79].

A single clinical study of ^{170}Tm–EDTMP in a 68 year old patient with metastatic prostate cancer was reported [257]. A freeze-dried kit for the preparation of ^{177}Lu–EDTMP was used for the preparation of the radiopharmaceutical [246]. The kit, containing 35 mg of EDTMP, 14.1 mg of NaOH and 5.8 mg of Ca^{2+}, was reconstituted in 1 mL of sterile saline and incubated with $^{170}TmCl_3$ at room temperature for 15 min. Figure 20 shows serial images of a patient injected with 370 MBq (10 mCi) of ^{170}Tm–EDTMP. The images were taken at 7 days (b), one month (c) and two months (d) post-administration, whereas (a) shows an image with ^{99m}Tc–MDP in the same patient prior to ^{170}Tm–EDTMP administration. It is evident from these images that the distribution of ^{170}Tm–EDTMP is concordant with the ^{99m}Tc–MDP uptake, both in physiological areas of distribution and at the sites of metastases. The images show high skeletal accumulation of ^{170}Tm–EDTMP with high tumour to normal bone uptake ratios and selective retention therein for a long period. The images also indicated the theranostic potential of the agent, whereby extended dosimetric studies can be carried out to optimize the absorbed dose delivered to the tumour for adequate and prolonged pain relief while limiting the associated bone marrow suppression.

Because it is a long lived radionuclide, the specific activity of ^{170}Tm used in the radiopharmaceutical is moderate at 6.36 GBq/mg (170 mCi/mg). The amount of thulium in 370 MBq (10 mCi) of ^{170}Tm activity will be less than 0.1 mg, even when the specific activity is only ~370 MBq/mg (100 mCi/mg). A detailed toxicity study of thulium and other rare earth elements such as terbium and ytterbium has been reported [258]. The LD_{50} value in mice is 485 (466.3–504.4) mg/kg for intraperitoneal administration and 6700 (6374.9–7041.7) mg/kg for oral administration. Hence, <0.1 mg of thulium injected as part of the radiopharmaceutical preparation is not likely to result in any adverse toxic effects in patients. Approximately 50% of the radiopharmaceutical injected is taken up by the bone, from where it is not released to circulation, as evident from Fig. 20, and the remainder is excreted through the renal system. Hence, the probability of free thulium circulating in the blood and thereby inducing any toxicity is low.

Thulium-170 complexes with several other polyaminopolyphosphonic acid ligands have also reported; ^{170}Tm DOTMP showed the best pharmacokinetics as a bone agent [59]. The main advantage of ^{170}Tm is that it can be prepared in large quantities with adequate specific activity. Being a long lived radionuclide, it can be shipped to distant locations and also retained for use as and when patients need it.

Thulium-170 radiopharmaceuticals have the potential to be used for metastatic bone pain palliation either alone or alongside short lived radionuclides such as ^{153}Sm or ^{177}Lu, where the short lived radionuclide provides immediate pain relief and sustained pain relief is obtained from the long lived ^{170}Tm. The combination radiopharmaceuticals ^{170}Tm–^{153}Sm or ^{170}Tm–^{177}Lu can be prepared by adding the required activities of the individual radionuclides to the same freeze-dried kit.

7.4. BONE PAIN PALLIATION AGENTS RADIOLABELLED USING BIFUNCTIONAL CHELATING AGENTS (BFCAs)

Most of the bone pain palliation radiopharmaceuticals discussed above were radiolabelled with metallic radionuclides directly, taking advantage of the

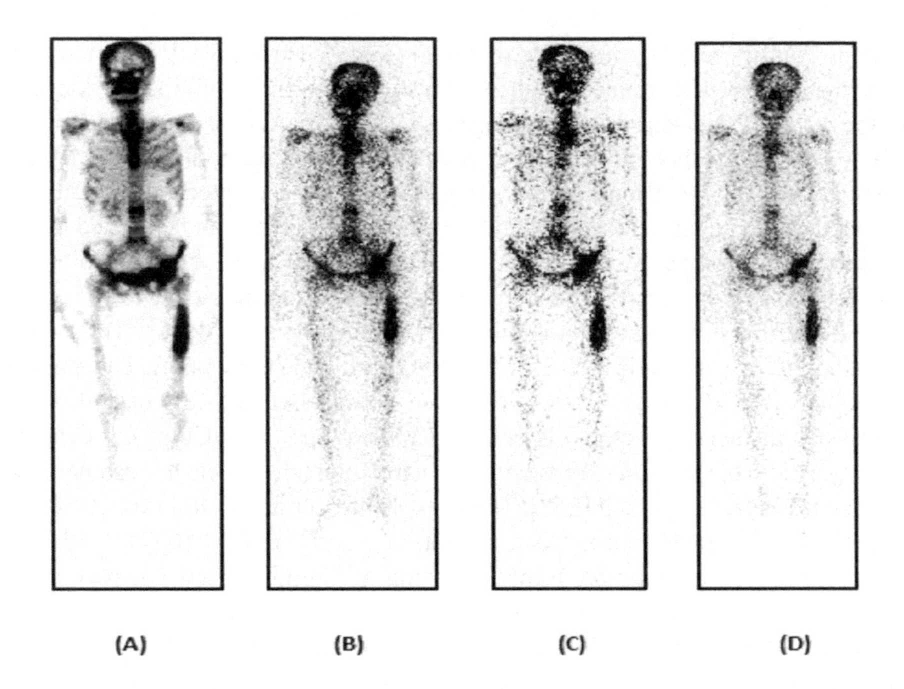

FIG. 20. A series of images representing whole body scintigraphic scans of a female patient suffering from metastatic breast cancer. (a) Diagnostic scan recorded with ^{99m}Tc–MDP. (b–d) Diagnostic scans for the same patient after the administration of 370 MBq (10 mCi) of ^{170}Tm–EDTMP at 1 week (b), 1 month (c) and 2 months (d) post-administration, respectively (courtesy of M.R.A. Pillai, Molecular Group of Companies).

complexation properties of the phosphate group present in the phosphonates. While this approach has been highly successful, an evolving strategy is to utilize the BFCA approach, in which the chelate is used to complex both diagnostic and therapeutic radionuclides, while a pendant bisphosphonate is conjugated to the BFCA for targeting.

Bisphosphonates have emerged as effective drugs for the treatment of various metabolic bone diseases, including bone metastases [15, 259, 260]. Bisphosphonates are biologically more stable chemical analogues compared to naturally occurring pyrophosphates. The labile P–O–P motif present in pyrophosphates is replaced by the more stable P–C–P unit in bisphosphonates (Fig. 21).

Bisphosphonates exhibit very high binding affinity for crystallized HA ($Ca_{10}(PO_4)_6(OH)_2$, Ca/P = 1.66), which is the main component of the inorganic matrix of bone. This high affinity is presumably owing to binding of the deprotonated phosphonate units with the Ca^{2+} present on the surface of crystalline

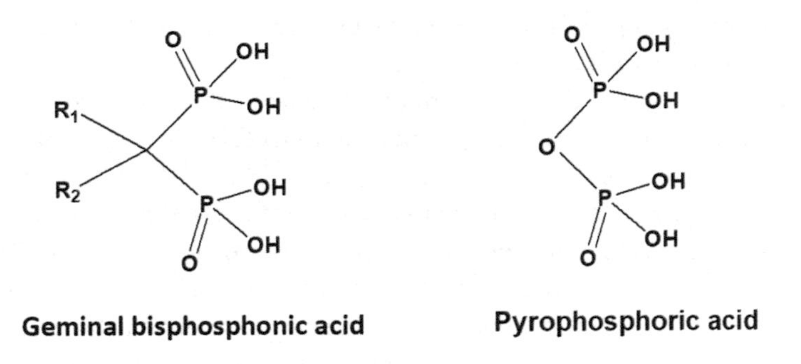

FIG. 21. Generic structures of geminal bisphosphonic acid and pyrophosphoric acid (courtesy of M.R.A. Pillai, Molecular Group of Companies).

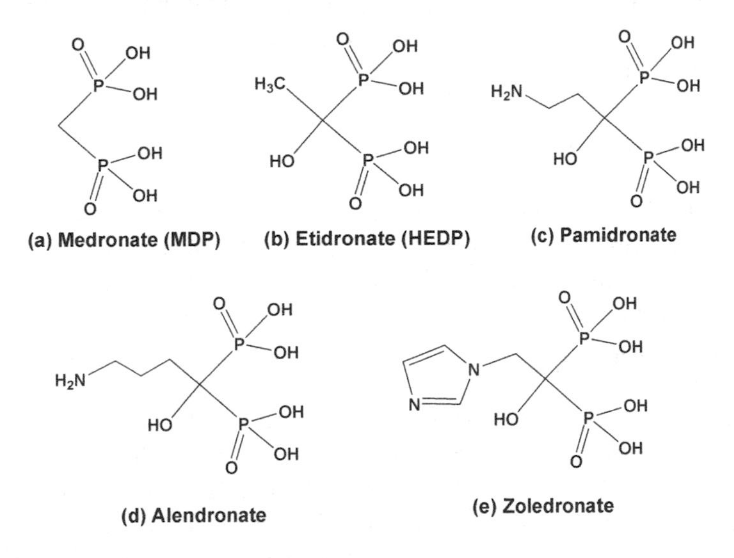

FIG. 22. Structures of the most common bisphosphonates synthesized and biologically evaluated (courtesy of M.R.A. Pillai, Molecular Group of Companies).

HA, likely in bidentate fashion [198]. In vitro and in vivo studies have shown that such binding of a bisphosphonate moiety to HA on the bone surface prevents both crystal growth and destruction of HA [198]. It is pertinent to note that when the R_2 substituent is modified for complexation with the radiometal, the nature of the R_1 substituent of the bisphosphonate moiety (Fig. 21) plays a major role in determining the potency of these molecules as drugs [32, 198, 261, 262]. The chemical structures of the most commonly synthesized and biologically evaluated bisphosphonates are shown in Fig. 22.

Management of painful skeletal metastases requires a multifaceted approach involving the use of systemic analgesic drugs, steroids, chemotherapeutic drugs, external beam radiation, radiofrequency ablation and therapeutic radiopharmaceuticals. Systemic internal radionuclide therapy using α^- and β^- emitting radionuclides, such as ^{223}Ra, ^{32}P, ^{89}Sr, ^{153}Sm, ^{186}Re and ^{177}Lu, has been proven to be the most effective modality with the fewest side effects, especially in the case of extensive multifocal osseous metastases [6, 137, 198, 204, 263]. Taking into consideration the high affinity and enhanced accumulation of the bisphosphonates to the bone matrix with abnormal osteoblast and osteoclast mediated activities, this class of molecules has been utilized effectively as molecular vectors for the delivery of radionuclides to the metastatic lesion sites.

The following section gives an account of the recent research and development on bisphosphonate ligands, which are modified with the addition of a BFCA to complex with radiometals as radiopharmaceuticals for palliative care of painful bone metastases.

7.4.1. ^{177}Lu–BPAMD

The first BFCA conjugated bisphosphonate ligand to be used successfully in the clinic was the bisphosphonate amide of DOTA, namely BPAMD ((4-{[(bis(phosphonomethyl))carbamoyl]methyl}-7,10-bis(carboxymethyl)-1,4,7,10-tetraaza-cyclododec-1-yl) acetic acid)). In order to develop this molecule, the bisphosphonate moiety is conjugated to DOTA through amide linkage (Fig. 23).

Initially, the ligand BPAMD was complexed with ^{68}Ga and the resultant radiotracer showed utility in terms of high bone accumulation in small animal PET experiments [264]. The ^{68}Ga–BPAMD complex also showed excellent pharmacokinetics and good target localization in clinical investigations [265]. The promising results obtained in diagnostic examinations using ^{68}Ga–BPAMD eventually led to the clinical application of ^{177}Lu–BPAMD. A comparative biodistribution study between ^{177}Lu–BPAMD and ^{177}Lu–EDTMP indicated higher skeletal accumulation as well as a higher target to background ratio for ^{177}Lu–BPAMD [266]. This is attributed to two factors, namely, superior in vivo kinetic stability of the Lu–BPAMD complex compared to its EDTMP counterpart and non-involvement of phosphonate groups in the radiometal chelation process, which renders them fully available for binding with Ca^{2+} present on the surface of mineralized bone. Due to its lower stability, a much larger quantity of ligand is required for the preparation of ^{177}Lu–EDTMP, which leads to potential target blocking. Preliminary human clinical investigations carried out using the ^{177}Lu–BPAMD formulation demonstrated that it has good potential as a radiopharmaceutical for bone targeted endo-radiotherapy [267]. Meckel et al.

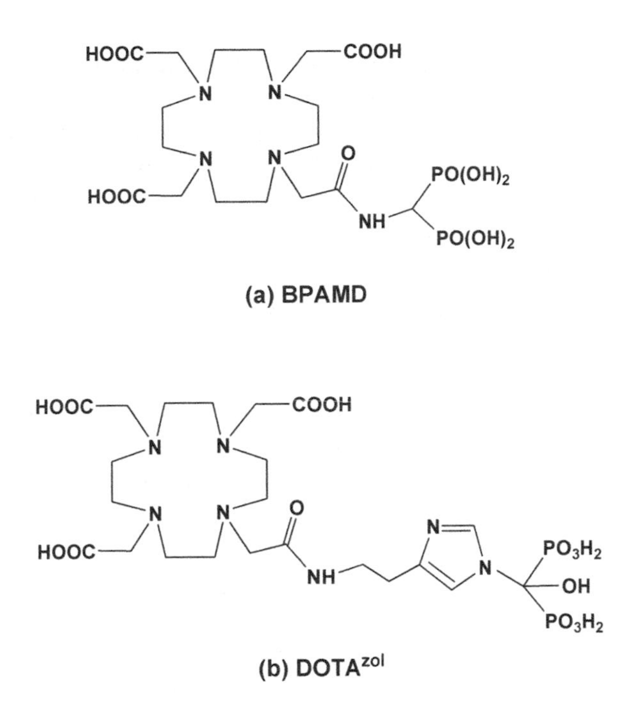

FIG. 23. Structures of DOTA coupled bisphosphonates: (a) BPAMD and (b) DOTA^{zol} (courtesy of M.R.A. Pillai, Molecular Group of Companies).

reported a kit type synthesis protocol and an automated radiolabelling module suitable for the routine formulation of the agent in a hospital radiopharmacy [268].

7.4.2. ^{177}Lu–DOTA^{zol}

Despite the promising clinical results obtained in palliative radiotherapy using ^{177}Lu–BPAMD, there remains further scope for improvement with respect to enhancing the accumulation of the radiotracer in metastatic bone lesions and simultaneous reduction of the uptake in the non-target organs. The side arms of the central carbon atom in the bisphosphonate structure play a crucial role in determining the skeletal uptake of radiolabelled bisphosphonates. It has been reported that when an –OH group (such as R_1 in Fig. 22(c)) is introduced in the bisphosphonate moiety, the affinity of the bisphosphonate towards hydroxyapatite is enhanced and consequently the accumulation of the agent in the skeleton is augmented [198]. Additionally, the availability of an aromatic nitrogen atom in the R_2 side chain helps in the formation of hydrogen bonds, which also eventually enhance the accumulation of the agent in bony metastases [269]. Zoledronic acid (Fig. 22(e)) is an example of one such bisphosphonate. These molecules have also been reported to influence the associated biochemical processes whereby they exert an inhibiting effect on the enzyme.

Keeping in mind the factors mentioned above, the potential of radiolabelled DOTA conjugated zoledronate (DOTA^{zol}, Fig. 21(b)) was evaluated as a bone targeting agent. The molecule was synthesized, radiolabelled with ^{68}Ga–^{177}Lu and subsequently evaluated by means of biodistribution as well as small animal PET studies. A comparison of biodistribution patterns of ^{68}Ga–DOTA^{Zol} with [^{18}F]NaF and ^{68}Ga–BPAMD showed that ^{68}Ga–DOTA^{Zol} has superior bone accumulation with minimal uptake in the soft tissues. On the other hand, biodistribution studies of ^{177}Lu–DOTA^{Zol} in normal healthy Wistar rats revealed high skeletal accumulation (3.43 ± 0.40%ID/g) and low uptake in the other non-target organs/tissues (blood: 0.07 ± 0.01%ID/g; muscle: 0.02 ± 0.00%ID/g) within 1 h of administration. These studies indicated that both ^{68}Ga and ^{177}Lu labelled DOTA^{Zol} have favourable in vivo properties, such as excellent accumulation in skeletons and desirable pharmacokinetics, which may enable the application of these two agents as a radiotheranostic pair for bone pain palliation.

Preliminary clinical investigations of ^{177}Lu–DOTA^{zol} have been carried out in four patients with bone metastases for pharmacokinetics and dosimetry studies [270]. Whole body scintigraphic images of the patients acquired at different post-administration time points showed excellent targeting of metastatic skeletal lesion sites, fast renal clearance and a high target to background ratio. Further, preliminary dosimetry studies have shown that ^{177}Lu–DOTA^{zol} also resulted in significantly lower bone marrow absorbed dose compared to ^{177}Lu–EDTMP, which in turn would allow the administration of a higher therapeutic dose of ^{177}Lu–DOTA^{zol} in patients as compared to ^{177}Lu–EDTMP.

7.4.3. ^{225}Ac–DOTA^{zol} for targeted alpha therapy

The LET (in the range of 100 keV/μm) and short tissue range (50 to 100 μm, covering approximately two to three cell diameters) of α particles are advantageous in a therapeutic context. Localized energy deposition is possible, while the energy deposited per unit mass of an α particle is nearly 100 times greater than that of a β^- particle [271]. The radiation resistance of cells caused by cross-fire effect of the β^- particle emitters is also reduced with α emitters [272].

Among the group of α emitting radionuclides used for in vivo targeted therapy, ^{225}Ac has been considered to be one of the most promising candidates due to its 10 day half-life and emission of four high energy α particles during its successive decay [273]. The first human clinical investigation using ^{225}Ac labelled PSMA inhibitor PSMA-617 in two prostate cancer patients was reported by Kratochwil et al. [65]. ^{177}Lu–PSMA-617 resistance was observed in one patient, but both patients showed a positive response to therapy. Subsequently, more elaborate clinical studies with ^{225}Ac–PSMA-617 have been performed in a larger group of patients with advanced stage, metastatic castration resistant prostate carcinoma with a PSMA positive tumour phenotype, with good clinical response, albeit with significant salivary gland uptake [274]. These advantages of ^{225}Ac targeted α therapy could also be utilized in alleviating pain in cancer patients with minimal side effects using macrocyclic bisphosphonates, which will not have salivary involvement.

Pfannkuchen et al. [275] were the first to report the synthesis, quality control and in vivo evaluation in animal models of ^{225}Ac–DOTA^{zol}. Biodistribution studies of ^{225}Ac–DOTA^{zol} in healthy Wistar rats showed high femur uptake (SUV = 4.86 ± 0.3 at 1 day and 4.99 ± 0.97 at 10 days post administration), which is comparable to ^{177}Lu–DOTA^{zol} (SUV = 4.18 ± 0.36 at 1 day and 4.63 ± 0.36 at 8 days post-administration). The authors documented that no clinical toxicity was observed in the animals up to two months post-treatment. These preliminary investigations showed the promise of ^{225}Ac–DOTA^{zol} as a future generation radiopharmaceutical for therapeutic use in painful bone metastases.

7.5. CONCLUSION

The development of radiopharmaceuticals based on ^{32}P and ^{89}Sr was a major advance in alleviating the pain of patients suffering from metastatic bone pain. Both these radiopharmaceuticals suffered from disadvantages. While ^{32}P could be made in large quantities for therapy, its high energy β^- particles induced bone marrow toxicity. Strontium-89 was very useful for therapy, with less bone marrow toxicity; however, its lack of availability is a drawback. The introduction of ^{153}Sm–EDTMP by the University of Missouri, Columbia, was a great step forward, using phosphonate ligands and radionuclides that can be prepared in large quantities using medium flux research reactors. Nevertheless, the short half-life of ^{153}Sm (47 h) did not allow its worldwide use.

The use of ^{177}Lu offers several advantages, such as the ability to be produced in large quantities, a reasonably long half-life of 6.73 days, lower bone marrow toxicity due to low energy β^- particles, and the ability to make stable complexes with both polyaminopolyphosphonates (e.g. EDTMP, DOTMP) and polyaminopolycarbonate ligands (e.g. DOTA). The haematotoxicity — which is the main limitation of radionuclide therapy — is much reduced in the case of ^{177}Lu, as the energy of β particles is sufficiently low and does not affect bone marrow tissue. The higher accumulation and retention of ^{177}Lu complexes in bone spares the other non-target organs from receiving unnecessary radiation dose, which adds to the therapeutic advantages of these bone pain palliative agents. The IAEA support for the clinical trials for ^{177}Lu–EDTMP was a milestone towards the approval of ^{177}Lu–EDTMP for use as a radiopharmaceutical for bone pain palliation in some Member States.

Although not many clinical studies have yet been reported, ^{170}Tm based radiopharmaceuticals for bone pain palliation offer a great opportunity. The radionuclide can be prepared in large quantities and the long half-life even allows for a stock of ^{170}Tm activity to be retained in nuclear medicine departments for the reconstitution of freeze-dried kits as and when a patient needs a therapeutic administration. A 'cocktail' therapy, or a mixed radionuclide therapy using ^{170}Tm and ^{177}Lu labelled phosphonates, may be advantageous to provide both short and long term pain relief from a single administration of a radiopharmaceutical. Adding the required amount of activity of both the radionuclides to a single freeze-dried kit of a phosphonate will be sufficient to make the two radiopharmaceuticals in the same vial.

Guerra Liberal et al. [68] carried out studies investigating the cellular effect of 11 different β^- particle emitting radionuclides: ^{32}P, ^{89}Sr, ^{90}Y, ^{117m}Sn, ^{153}Sm, ^{166}Ho, ^{170}Tm, ^{177}Lu, ^{186}Re and ^{188}Re. The computational models they used were able to estimate DNA damage and the probability of repair of DNA was accurately predicted. Their results showed maximum cell death when cells were exposed to ^{170}Tm and ^{177}Lu [68]. The results of this study are significant in enhancing the use of these two radionuclides for bone pain palliation in addition to α particle emitting radionuclides.

The latest development of coupling BFCA molecules to geminal bisphosphonates is also a significant addition to the ongoing research and development for bone seeking radiopharmaceuticals. The phosphonate groups are involved in binding with the calcium HA in the bone matrix and the release of these groups from coordination with radiometals is an intelligent approach. The 'zoledronate' based ^{177}Lu–DOTA^{zol} is found to be more potent, with improved pharmacological properties and higher accumulation in bone lesions in preliminary clinical studies. Replacing the β^- emitter ^{177}Lu with the α emitter ^{225}Ac led to the development of ^{225}Ac–DOTA^{zol}, with the potential for targeted α therapy of skeletal metastases with almost no bone marrow toxicity.

8. TARGET SPECIFIC RADIOPHARMACEUTICALS (TSRs)

8.1. TARGETED RADIONUCLIDE THERAPY

For the last two decades, most of the research in the radiopharmaceutical sciences has been focused on biomolecular chemistry. Consequently, there is extensive knowledge related to the chemical, biochemical and biokinetic behaviour of TSRs, which have been positioned as one of the best therapeutic options for certain advanced diseases [2, 99, 276].

TSRs are unique in their ability to detect characteristic phenotypes in vivo, when cells overexpress distinctive receptor proteins involved in the pathology. In general, therapeutic radiopharmaceuticals are composed of a metallic radionuclide, a BCFA and a bioactive fragment [276]. Specific bioactive molecules or biological vectors used as target seeking pharmacophores include peptides, enzyme inhibitors, aptamers, humanized or chimeric antibodies, antibody fragments, peptidomimetics, non-peptide receptor ligands and DNA analogues. The molecular weight influences the in vivo pharmacokinetics and it is expected that clearance from the blood and non-target organs and accumulation in the target tissue of interest is faster for small biomolecules, such as organic drugs and peptides, as compared to larger molecular weight proteins and antibodies. Hence, radiopharmaceuticals that are based on small carrier molecules are likely to provide better contrast in diagnostic images and deliver a lower radiation dose to non-target tissues in the case of therapeutic products.

Figure 24 illustrates the general approach for developing TSRs [277]. The identification of the suitable target and validation of the presence of that target for the disease of interest is the first step for any drug or radiopharmaceutical development. Cancer cells are now known to express various proteins at much higher magnitudes than their original normal cell type. Several such receptor proteins serve as a target and a large number of ligands or synthetic molecules capable of binding to such targets are identified, synthesized and screened, and selected ones are further taken for drug development. For radiopharmaceutical development, these selected target seeking molecules are modified through the addition of a suitable BFCA molecule, as metallic radionuclides are generally used for diagnosis (^{68}Ga or ^{99m}Tc) or therapy (^{90}Y, ^{177}Lu or ^{225}Ac) [278]. A spacer group encompassing a carbon backbone and at times modified with aromatic rings or hydrophilic groups (e.g. alcohols) is sometimes introduced in the design of a radiopharmaceutical to conserve the biological activity of the carrier molecule by distancing it from the radionuclide chelating moieties. The choice of different spacer groups allows modulation of the lipophilicity parameters, when necessary [277].

(^{68}Ga or ^{99m}Tc) or therapy (^{90}Y, ^{177}Lu or ^{225}Ac) [278]. A spacer group encompassing a carbon backbone and at times

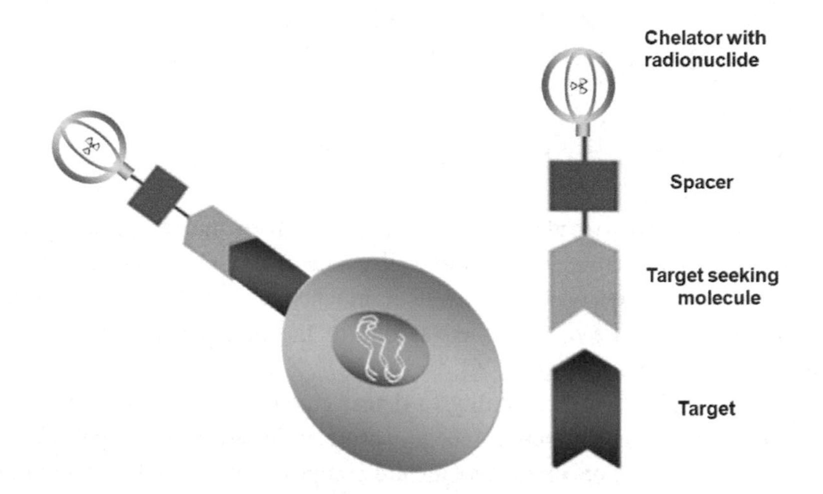

FIG. 24. An illustration of a TSR. The target is a cellular molecule that is commonly present on the surface of cancer cells in higher numbers than on that of normal cells. A target seeking molecule with high affinity is attached to a radionuclide via chemical modifications. The radionuclide is either a diagnostic one (^{68}Ga or ^{99m}Tc) or a therapeutic one (^{90}Y, ^{177}Lu or ^{225}Ac) and forms a stable complex with BFCA; in order not to lose molecular recognition, a spacer is introduced between the target seeking molecule and the BFCA that carries the radionuclide (reproduced from Ref. [277] with permission).

modified with aromatic rings or hydrophilic groups (e.g. alcohols) is sometimes introduced in the design of a radiopharmaceutical to conserve the biological activity of the carrier molecule by distancing it from the radionuclide chelating moieties. The choice of different spacer groups allows modulation of the lipophilicity parameters, when necessary [277].

8.2. DESIRABLE FEATURES OF TARGET SEEKING RADIOPHARMACEUTICALS FOR THERAPY

Ideally, a TSR used for therapy ought to not only alleviate pain in patients with advanced cancer, but also induce a significant survival rate. For this purpose, a destructive effect on malignant cells ought to be induced without exceeding the maximum tolerated absorbed dose to bone marrow and other critical organs. The following characteristics are desirable for a TSR:

(a) High affinity for the target molecules overexpressed in cancer cells;
(b) Fast elimination from non-target organs and tissues;

(c) High in vivo stability;
(d) High internalization rate in cancer cells;
(e) High LET radiation to induce DNA double strand breaks;
(f) Short range particulate radiation for destruction of tumour cells with low damage to surrounding normal tissue;
(g) Ablative effect independent of cellular oxygenation in order to eliminate hypoxic tumour cells.

8.3. TARGET SPECIFIC RADIOPHARMACEUTICALS FOR PROSTATE CANCER

Prostate and many other types of cancers overexpress an enzyme known by multiple names, such as carboxypeptidase (GCP II), *N*-acetyl-L-aspartyl-L-glutamate peptidase I, *N*-acetyl-L-aspartyl-L-glutamate peptidase and prostate specific membrane antigen (PSMA) [279]. The overexpression of PSMA in prostate cancer cells has been reported [279, 280]. PSMA was used as a target for the development of TSRs for the theragnosis of prostate cancer using monoclonal antibodies [62, 281, 282]. Although most authors have referred to the enzyme as PSMA, it is expressed in several other cancers other than that of the prostate.

8.3.1. PSMA enzyme and its action

Enzymes are proteins and many enzymes contain metal ions that are known as metalloenzymes; PSMA is one such metalloenzyme protein with zinc, having a molecular weight of 100–104 kDa [283]. The protein structure of PSMA shows an arrangement of 750 amino acids, in which 24 amino acids are located in the cell membrane site, 19 are intracellular and the major part (707 amino acids) are found in the extracellular portion (Fig. 25). Most of the enzyme resides in the extracellular space and is not secreted to the bloodstream [284–287]. As PSMA is overexpressed in several types of cancers, it is used as a target to develop radiopharmaceuticals for the diagnosis and therapy of such cancers [277].

PSMA is shown to have enzyme activity like other zinc metallopeptidases. PSMA uses different substrate molecules in different tissues, but its main role is linked to cellular folate (vitamin B9) levels. In the intestines poly-gamma glutamate folate acts as a substrate and gamma linked glutamate is removed by PSMA, making folate available for cellular activity. Overexpression of PSMA on many tumour tissues is thought to be linked with this mechanism. PSMA converts the poly-gamma glutamate folate from dying tumour cells and folate is made available to healthy cells in the vicinity, which

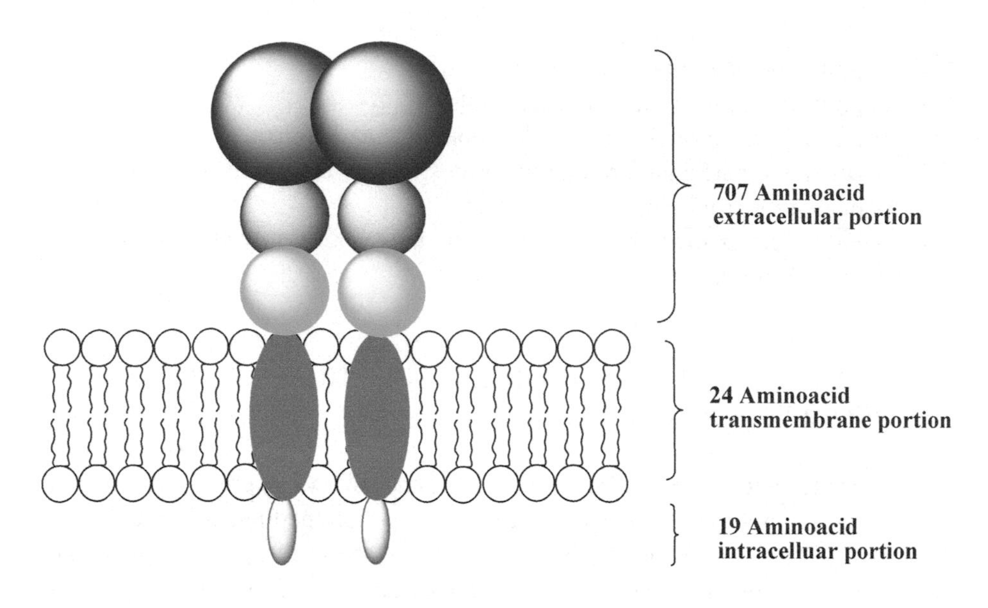

FIG. 25. Schematic representation of PSMA enzyme that consists of an extracellular portion of 707 amino acids, a transmembrane portion of 24 amino acids and an intracellular portion of 19 amino acids (reproduced from Ref. [286] with permission).

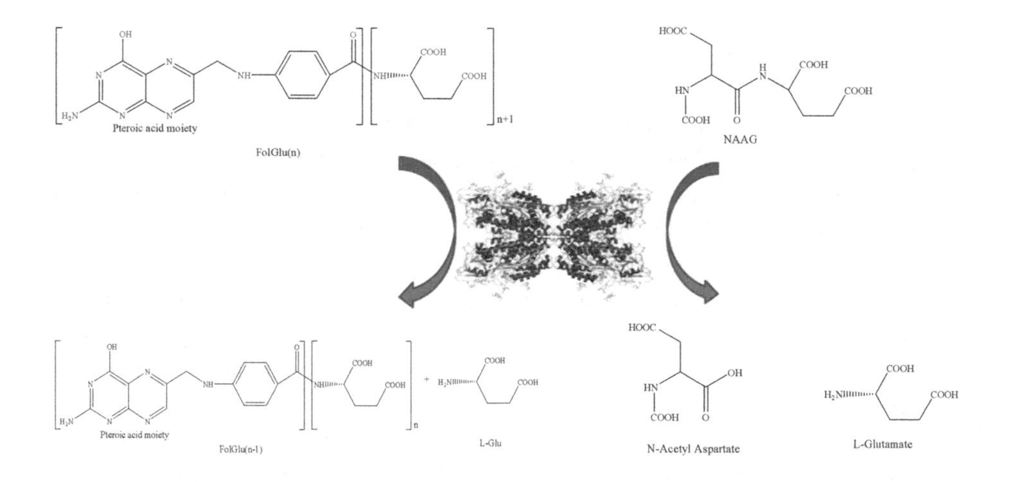

FIG. 26. Two different enzyme actions of PSMA. Glutamic acid is released from folate polyglutamate, resulting in the release of folic acid, and after successive releases of glutamate, folate is released, which is a B vitamin (right). N*-acetyl-*L*-aspartyl-*L*-glutamate is hydrolyzed to aspartate and glutamate (left).*

triggers excessive proliferation and tumour growth. The PSMA also acts by hydrolyzing *N*-acetyl-L-aspartyl-L-glutamate, where aspartate and glutamate are produced (Fig. 26).

With enzymes such as PSMA being proteins, they can be targeted using specific monoclonal antibodies, which was the basis for the development of the first approved target specific radiopharmaceutical using a BFCA, namely prostascint [288]. Efforts to develop therapeutic radiopharmaceuticals using monoclonal antibodies had very limited success, mainly because of the large molecular weight of the antibodies or fragments and consequently the time taken by them to accumulate in the target.

The action of an enzyme is initiated by binding to its specific substrates. Molecules that mimic these substrates can bind to enzymes, but they do not trigger further enzymatic signalling actions. Several drugs act by this mechanism of enzyme inhibition. The anticancer drug methotrexate is one such commonly used enzyme inhibitor of dihydrofolate reductase that inhibits nucleotide synthesis and prevents further cell proliferation [289]. PSMA is a valid target for prostate and some other types of cancer and several molecules with potential inhibitory action against the enzymatic activity of PSMA have been reported and reviewed in the literature [290, 291].

8.3.2. TSRs based on the enzyme activity of PSMA and prostate cancer imaging

The use of 'inhibitor' ligands targeting PSMA as a target molecule is a major development in radiopharmaceutical chemistry. The radiopharmaceutical ^{68}Ga–PSMA-11 (^{68}Ga labelled Glu-NH-CO-NH-Lys-(Ahx)-HBED-CC) (Fig. 27), first reported for PET–CT imaging of prostate cancer in 2013, is the culmination of a series of research initiatives concerning the development of radiopharmaceuticals with enzyme inhibitors [292–297]. ^{68}Ga–PSMA-11 is a successful tracer used for imaging not only prostate cancer but several other types of neoplasm, such as hepatocarcinoma, glioma, clear cell renal cell carcinoma, papillary carcinoma of the thyroid and adenocarcinoma of the intestine [277].

PSMA-11 uses the ligand Glu-NH-CO-NH-Lys (Glu-Urea-Lys), a PSMA enzyme inhibitor molecule, which is conjugated with the chelating agent HBED–CC (*N,N′*-bis[2-hydroxy-5-(carboxyethyl)benzyl] ethylenediamine-*N,N′*-diacetic acid) at the lysine end through a spacer molecule [296]. In the Glu-Urea-Lys/PSMA interaction, three −COOH groups of Glu and Lys in the ligand interact electrostatically with peptides located in the enzyme active centre, and the oxygen is coordinated to zinc ions. However, in the Glu-Urea-Lys molecule, an additional lipophilic site is added to facilitate the reaction with the hydrophobic central pocket of the enzyme. Figure 28 shows

PET–CT images of patients with (a) suspected disease, (b) a single lesion in the iliac bone and (c) extensive soft tissue and bone metastases [277]. The images have excellent contrast, with very high tumour to background ratios. The activity seen in the salivary and lachrymal glands is a normal uptake pattern for this tracer.

8.3.3. PSMA inhibitor ligands for therapy of prostate cancer

8.3.3.1. ^{177}Lu–PSMA therapy

HBED-CC is an amine phenol ligand with very high affinity for ^{68}Ga and thus forms stable complexes with very small amounts of ligand [298]. However, it is not useful as a chelating agent for any of the lanthanide radiometals. Lutetium is known to form stable complexes with the DOTA ligand, which is conjugated to the pharmacophore Lys–Urea–Glu for developing ^{177}Lu based therapeutic radiopharmaceuticals. DOTA is a hydrophilic chelating agent and spacer molecules are synthesized with benzene, pyridine or cyclohexane rings to increase the lipophilicity needed for enzyme interactions. The therapeutic application of three different ^{177}Lu labelled PSMA inhibitors (^{177}Lu–PSMA–617, ^{177}Lu–PSMA–I&T and ^{177}Lu–iPSMA) (Fig. 29) has been reported [299–301]. All three molecules have the same pharmacophore (Lys–Urea–Glu) and BFCA (DOTA) and differ only in the spacer molecules used. Among the three PSMA ligands mentioned, PSMA-617 is the most widely used for therapy, as it is commercially available (Fig. 29(a)).

PSMA contains two zinc ions in its active centre. In the Glu–Urea–Lys/PSMA interaction, the three −COOH groups of Glu and Lys in the ligand interact electrostatically with the amino acids located in the enzyme active centre, and the oxygen from the carbonyl is coordinated to zinc ions [284]. The lipophilic sites

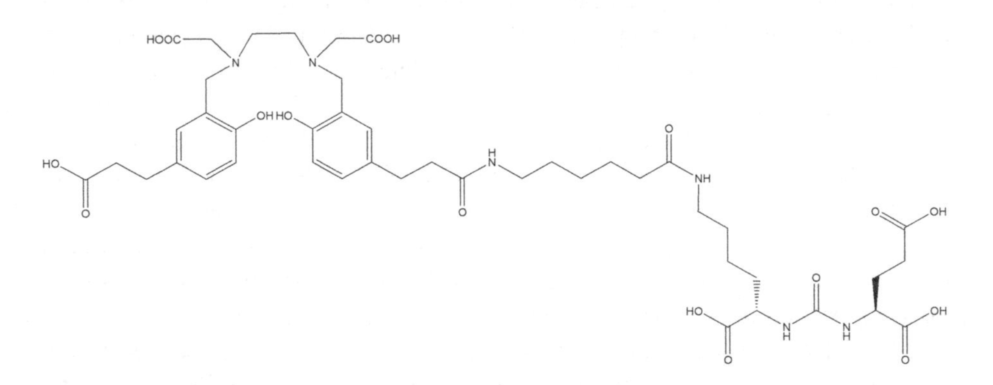

FIG. 27. PSMA-11 ligand (Glu-CO-Lys(Ahx)-HBED-CC or Glu-NH-CO-NH-Lys(Ahx)-HBED-CC) (courtesy of M.R.A. Pillai, Molecular Group of Companies).

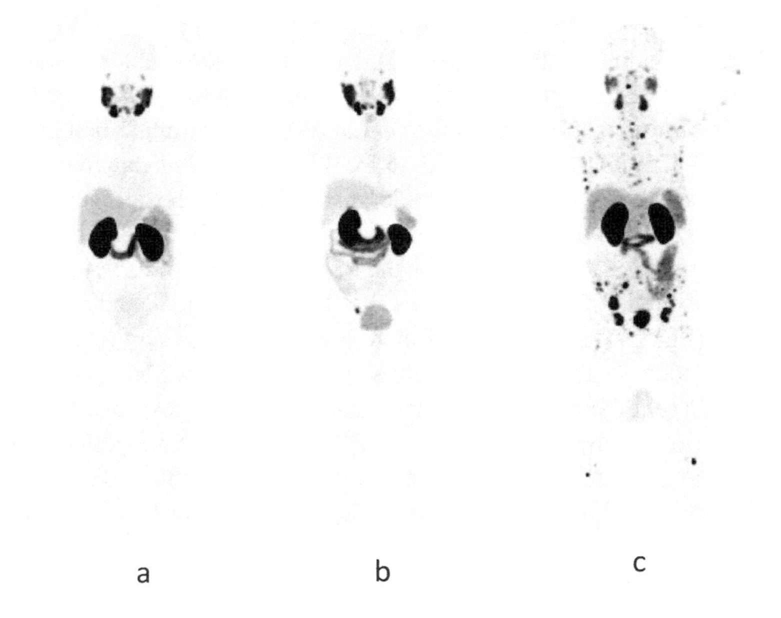

FIG. 28. [^{68}Ga]PSMA-11 PET–CT imaging performed 1 h after administration in three patients. (a) A patient suspected to have prostate cancer; hard nodule in right lobe of prostate on digital rectal examination, however no disease detected in PET–CT imaging. No prostate cancer was detected. Uptake seen in kidneys, salivary glands, parotid glands and lachrymal glands. (b) A 76 year old prostate cancer patient having undergone orchidectomy and on hormonal therapy. A single lesion is seen in the iliac bone. (c) A 45 year old patient diagnosed with adenocarcinoma of the prostate with soft tissue and extensive bone metastasis (reproduced from Ref. [277] with permission).

introduced in the PSMA ligands are necessary for coupling with the hydrophobic central pocket of the enzyme. The probable sites where substrate and inhibitors bind to the enzyme are predicted from the enzyme structure, which is a tunnel like structure ~20 Å deep from the surface of the enzyme [287]. Apparently, this is the reason for introducing a long spacer molecule between the pharmacophore (Glu–Urea–Lys) and the chelate. The aliphatic and aromatic groups added to the Glu–Urea–Lys molecule account for the differences in chemical properties among the ^{177}Lu–PSMA inhibitors reported so far (Table 3).

Since the minimum energies associated with optimal structural geometry reveal the stability of a molecule (a high value of steric energy is an indication of instability), theoretical calculations (BioMedCAChe Work System) showed that Lu–PSMA–I&T and Lu–iPSMA molecules are more stable than Lu–PSMA-617 (Table 3). However, as Table 3 also demonstrates, the order of magnitude of the related energy in the most stable conformers is inverse to their ClogP values (Lu–PSMA-617 > Lu–iPSMA > Lu–PSMA–I&T), which suggests that Lu–PSMA-617 and Lu–iPSMA could have better biological properties than Lu–PSMA–I&T for coupling to the hydrophobic central pocket of the PSMA enzyme. The lower number of conformers and the higher angle bending energy values for Lu–iPSMA when compared to Lu–PSMA-617 indicate that HYNIC in Lu–iPSMA possibly acts as a more rigid spacer than the hexane in Lu–PSMA-617, promoting a better spatial orientation of the Glu–NH–CO–NH–Lys-(Nal) fragment for biological interaction with the PSMA active centre. Nevertheless, the clinical applications of ^{177}Lu–PSMA-617, ^{177}Lu–PSMA–I&T and ^{177}Lu–iPSMA radiotracers (Fig. 29) have shown that there is apparently no significant difference among their therapeutic efficacies. This result is as expected, since the main effect for pain relief and reduction of metastatic lesions is governed by ^{177}Lu.

Figure 30 presents ^{68}Ga–PSMA-11 PET–CT images for a patient with bone metastases (a) prior to therapy and (b) two months after the administration of ^{177}Lu–iPSMA (5.55 GBq or 150 mCi). A decrease in metastatic tumour masses

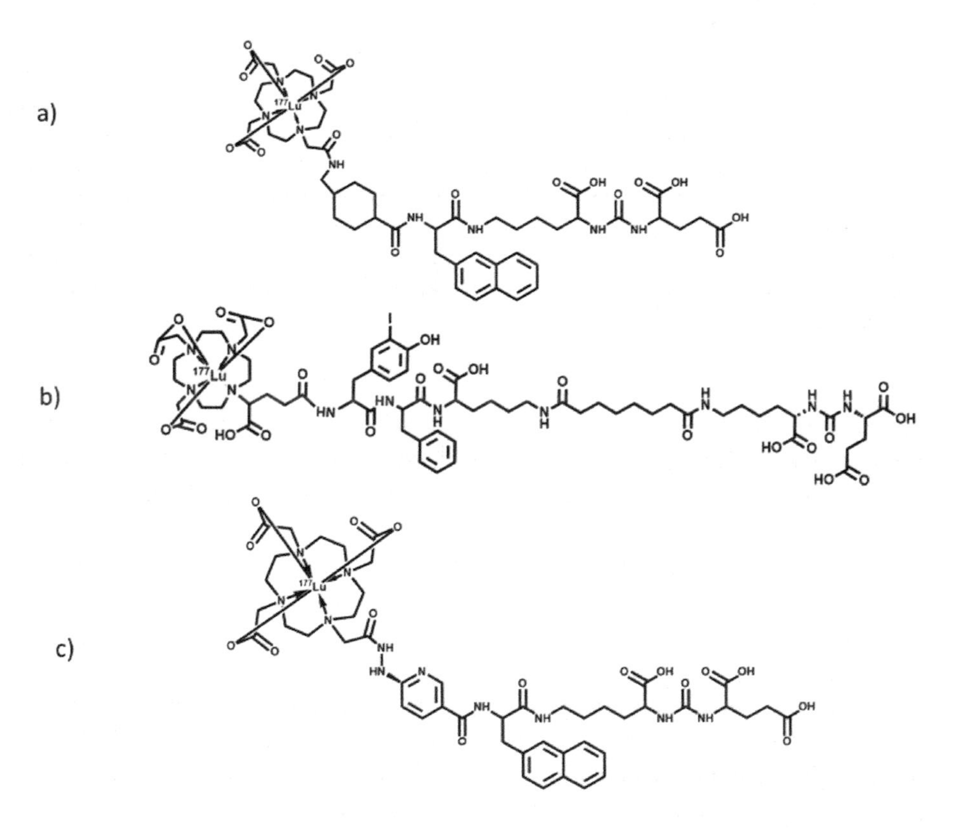

FIG. 29. Chemical structures of (a) ^{177}Lu–PSMA-617, (b) ^{177}Lu–PSMA–I&T and (c) ^{177}Lu-iPSMA (courtesy of G. Ferro-Flores, Instituto Nacional de Investigaciones Nucleares).

post-therapy can be visualized, which also resulted in complete pain relief for the patient. The patient was therefore prescribed a few more administrations of the radiopharmaceutical in order to improve quality of life and prolong survival.

In general, targeted radionuclide therapy with all three ^{177}Lu–PSMA inhibitors discussed above is well tolerated, safe and provides a significant

TABLE 3. CHEMICAL PROPERTIES OF DIFFERENT PSMA INHIBITOR LIGANDS

	Lu–PSMA-617	Lu–PSMA I&T	Lu–iPSMA
CLogP: (kcal/mol)	2.4634	CLogP: 2.1342	CLogP: 2.26716
Energy	−33.7	−81.0	−76.4
Electrostatic charge	27.1	35.5	19.9
Hydrogen bond	−197.9	−210.8	−206.8
Stretch	10.2	50.1	26.9
Angle	36.2	143.6	84.6
Torsion stretch	−1.9	−6.2	−3.269
Bend	0.8	8.5	3.044
Energy of the most stable conformer	−48.2	−275.0	−76.241
Number of conformers	46	9	24

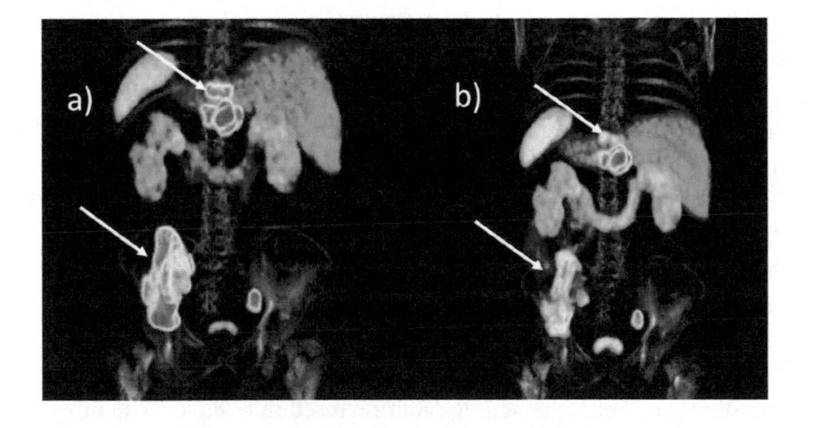

FIG. 30. (a) ^{68}Ga–PSMA-11 PET–CT image for a patient with bone metastases and unbearable pain. (b) ^{68}Ga–PSMA-11 PET–CT image for the same patient two months after the administration of ^{177}Lu–iPSMA (5.55 GBq, 150 mCi) (courtesy of G. Ferro Flores Instituto Nacional de Investigaciones Nucleares).

reduction in PSMA levels, as well as a decrease in metastatic progression and a pain relief response in patients with advanced prostate cancer. However, more randomized trials to evaluate the therapeutic effect of PSMA inhibitor based ^{177}Lu radiopharmaceuticals on patient survival are needed.

8.3.3.2. ^{225}Ac–PSMA therapy

It is well known that the deposition of DNA damaging energy by α particles is ~100 times greater than that by β particles [302]. Radiolabelling of PSMA ligands with α emitting radionuclides is an interesting proposition for developing TSRs. Targeted α therapy with ^{225}Ac–PSMA-617 has shown impressive results with significant benefits for patients with advanced stage prostate cancer [65, 303]. Figure 31 shows a ^{68}Ga–PSMA PET image of a patient with metastatic prostate cancer before (left) and after (right) treatment with ^{225}Ac–PSMA. Actinium therapy was shown to be particularly effective against visceral metastases, as seen in the case of this patient. The superior efficacy of ^{225}Ac therapy can be attributed to the high LET of the four α particles emitted by actinium and its three progenies during the radionuclide decay chain. Hence, a single ^{225}Ac decaying inside a cancer cell is equivalent to around 400–500 ^{177}Lu atoms in terms of energy delivered. In addition, only a part of the energy of β^- particles emitted by ^{177}Lu is received in a single cell due to its relatively long maximum path length of 1–2 mm.

The choice of radionuclide between ^{177}Lu and ^{225}Ac is currently linked to the severity of the disease, the availability of the radionuclide and affordability. ^{177}Lu–PSMA has a favourable dosimetry and a convincing response in terms of decrease in serum PSA level and RECIST (response evaluation criteria for reduction of the tumour mass in solid tumours) data. Nevertheless, ~30% of patients do not respond to ^{177}Lu labelled PSMA, despite good tolerance. Patients with a greater metastatic disease can benefit more from ^{225}Ac–PSMA. For the patient in Fig. 32(b), ^{177}Lu therapy is considered to be adequate, while the patient Fig. 32(a) with extensive bone metastasis is expected to benefit more from ^{225}Ac–PSMA therapy.

The undesirable effects of targeted therapy with radiolabelled PSMA ligands are mainly related to xerostomia, due to the salivary fixation of the tracer. The toxicity of salivary and lachrymal glands is expected to be more severe with ^{225}Ac than with ^{177}Lu due to the high LET and consequently large energy deposition of ^{225}Ac. Thus, the activity administered may have to be limited based on this undesirable side effect.

Currently, new PSMA inhibitors with better lipophilic properties are under development for labelling with ^{225}Ac or ^{177}Lu. Selective uptake of a radioligand by cancer cells is sufficient to obtain images that are adequate for diagnosis.

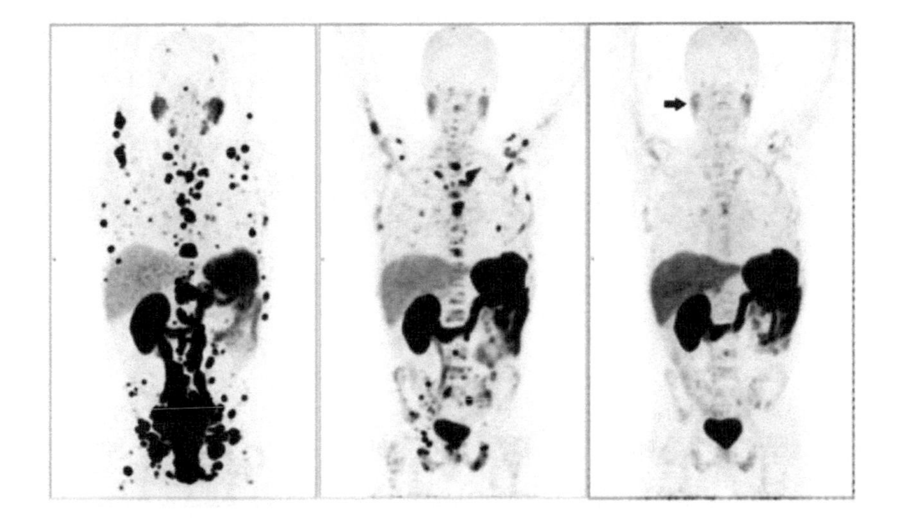

FIG. 31. ^{68}Ga–PSMA PET images of a patient with metastatic prostate cancer, before (left), after three sessions (centre) and at the end of treatment (right) with ^{225}Ac–PSMA-617. Maximum intensity projection (MIP) images present a progressive response in virtually all target lesions. Note also the diminishing physiological uptake in parotid glands (black arrow) (reproduced from Ref. [304] with permission).

However, to provide efficient therapy, a high rate of radiopharmaceutical internalization is necessary in order to delay the exposure of cancer cells to alpha or beta particle emissions during radionuclide disintegration [305]. This point is critical in the development of new ^{225}Ac–PSMA inhibitors, in which the recoil energy of the ^{225}Ac nucleus during decay can cause the release of the decay radionuclide and its progeny (^{221}Fr, ^{217}At, ^{213}Bi, ^{213}Po, ^{209}Tl, ^{209}Po) due to the breaking of the chemical bonds [302]. Other approaches for improving the efficacy of therapy are the encapsulation of ^{225}Ac in target specific nanoparticles and the development of new chelators that form highly stable actinium complexes that are internalized [306].

Although $^{223}RaCl_2$ is not a TSR, it is an alpha emitter (half-life of 11.43 days) approved for human use for the treatment of prostate cancer induced bone metastases [307]. $^{223}RaCl_2$ is taken up by bone with accumulation in metastases due to its calcium mimetic behaviour, while ^{225}Ac–PSMA-617 is taken up by cancer cells because it is recognized by the PSMA enzyme present in them. Even though the two radiopharmaceuticals are alpha emitters, with the same number of alpha particle emissions (four) and high LET, more damage to cancer cells at the same radiation absorbed dose in tumours is expected from ^{225}Ac–PSMA-617 than from $^{223}RaCl_2$ because ^{225}Ac remains within the cancer cells, while ^{223}Ra stays in the bone compartment. However, the relative uptake of the two

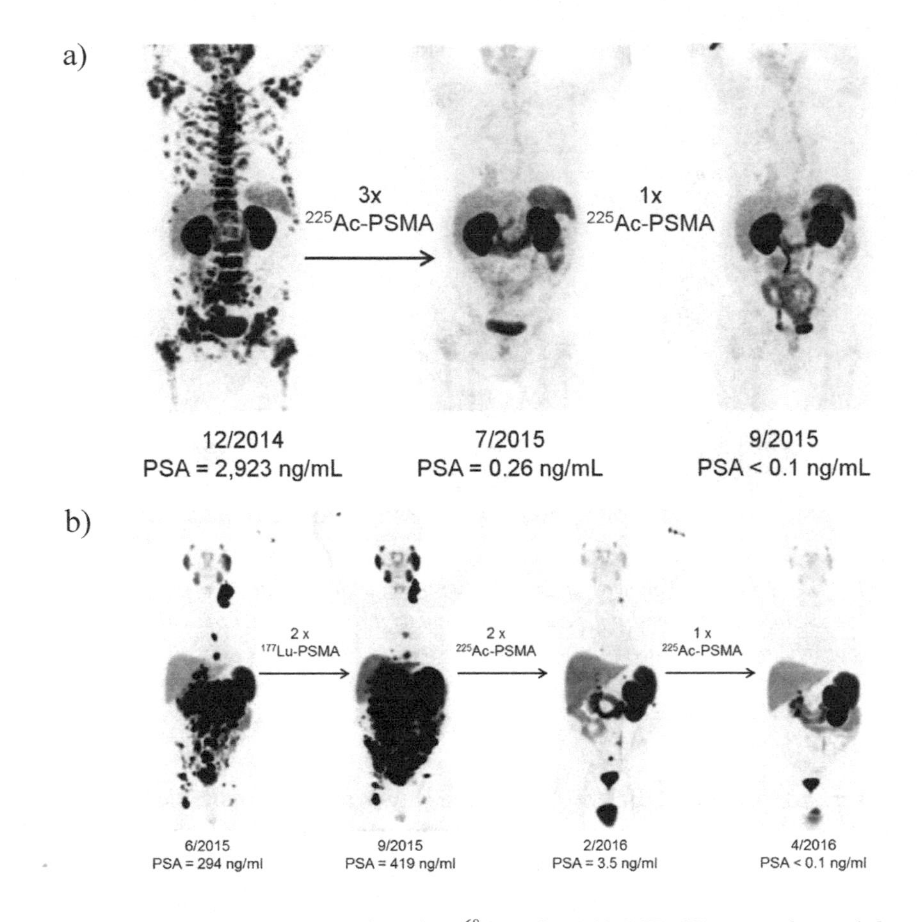

FIG. 32. Therapeutic strategy is based on ^{68}Ga–PSMA-11 PET–CT scans obtained during diagnosis. The patient in panel (a) is considered to be suitable for treatment with ^{225}Ac–PSMA therapy, whereas for the patient in panel (b) treatment with ^{177}Lu–PSMA is preferred (reproduced from Ref. [65] with permission).

radiopharmaceuticals by metastatic bone could vary and thus the comparative benefit needs to be estimated.

It is important to note a problem related to the nucleus recoil energy of ^{225}Ac–PSMA-617, which can cause the release of the decay radionuclides from the PSMA-617 molecule. However, the decay of the first two radionuclides, ^{221}Fr ($T_{1/2}$ = 4.9 min) and ^{217}At ($T_{1/2}$ = 32 ms), occurs within a few minutes; only the disintegration of ^{213}Bi ($T_{1/2}$ = 47 min) has a longer half-life and takes place at a later stage. It is highly probable that all the ^{225}Ac progenies will remain in the bone tissues once the TSR is internalized. Even if the progeny is released, it is likely to be close to the target cancer cells from which it was released. However, detailed studies are needed to assess these assumptions.

8.3.4. Synthesis of PMSA ligands

In general, Glu–Urea–Lys is synthesized through a reaction between the di-*tert*-butyl ester of glutamic acid and carbonyl diimidazole (CDI) to form the acyl imidazole derivative, which reacts with the (S)-*tert*-butyl-2-amino-6-(benzyloxycarbonylamine) hexanoate (Cbz-Lys-Ot-Bu); after Cbz deprotection, Glu–Urea–Lys is obtained. Thereafter, several modifications can be made; for example, the addition of β-naphthyl alanine, followed by the incorporation of hydrazinepyridin-3-carboxylic acid (HYNIC) or cyclohexane, and the DOTA chelator [301, 308]. Finally, chemical characterization is carried out by ^{1}H NMR, IR and mass spectroscopies.

8.4. TSRs BASED ON CHEMOKINE RECEPTOR-4 (CXCR-4) LIGANDS

Another interesting approach for the preparation of TSRs is the development of DOTA–CXCR4 (chemokine-4) ligands that bind specifically with chemokine receptors on cancer cells [309–311]. Chemokine receptors are G protein coupled receptors that mediate chemotaxis. Chemotaxis is the movement of an organism in response to a chemical stimulus. The chemokine receptor subtype CXCR4 triggers its biological effect by binding to the CXCL12 ligand. CXCR4 is overexpressed in more than 70% of human solid tumours, including breast and prostate cancer, cervical adenocarcinoma, B-cell lymphoma, melanoma, glioma and neuroblastoma, among others [3]. CXCR4 is involved in primary tumour growth, cancer cell migration and metastatic sites.

Cyclic pentapeptides have been successfully designed as ligands for CXCR4 [312]. In the sequence cyclo(D-Tyr-[*N*Me]-D-Orn(DOTA-4-aminomethylbenzoyl)-Arg-2-Nal-Gly) (pentixafor), the side chains of arginine and ornithine, as well as the methyl groups, contribute significantly to the binding affinity (Fig. 33). This derivative can be labelled with the theranostic ^{68}Ga–^{177}Lu pair. The development of new CXCR4 radioligands to be labelled with beta and alpha emitters is an important field of research into obtaining efficient TSRs for the management of primary cancer as well as palliation of bone pain due to metastases. These radiopharmaceuticals will also be useful for the management of prostate cancer [313]. Radiolabelled CXCR4 targeting ligands are an interesting addition to the armamentarium of peptide receptor radionuclide therapy [314].

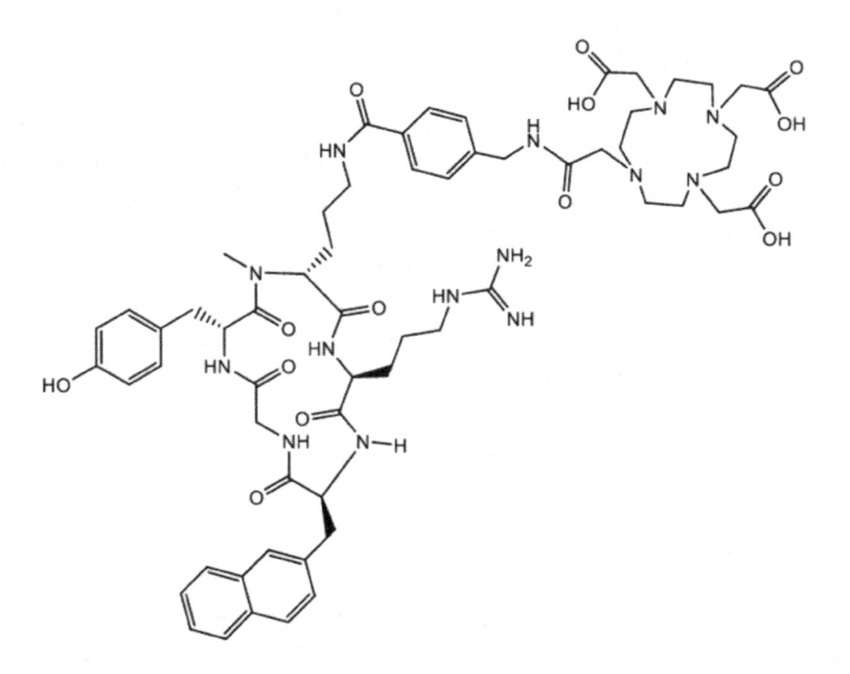

FIG. 33. The pentaxifor ligand, cyclo(D-Tyr-[NMe]-D-Orn(DOTA-4-aminomethylbenzoyl)-Arg-2-Nal-Gly), used for radiolabelling with metallic radionuclides for the development of theranostic radiopharmaceuticals.

8.5. TSRs BASED ON FAP INHIBITOR LIGANDS

The use of PSMA ligands was a significant step in radiopharmaceutical development as it was the first time an inhibitor molecule had been used as a pharmacophore in a radiopharmaceutical. A rigorous search for more such inhibitor molecules that are suitable for the development of TSRs is likely to continue. The use of fibroblast activation protein (FAP) inhibitors as the pharmacophore in the development of theranostic radiopharmaceuticals is becoming another success story [315].

It has been reported that FAP expression promotes tumour growth. FAP is a type II serine protease enzyme that cleaves peptides located after proline residues [316]. Inhibition of the FAP enzyme activity can help to increase the antitumour immune response. The FAP enzyme is overexpressed on the cell surface of the activated stromal fibroblasts present in most human epithelial tumours, but not in normal fibroblasts [316]. Hence, developing TSRs using FAP inhibitor (FAPI) ligands is an important strategy. Like PSMA inhibitor ligands, FAPIs are small molecules that can be modified suitably with a BFCA for radiolabelling with diagnostic and therapeutic radionuclides.

Jansen et al. reported the synthesis of 67 novel FAPIs to explore the structure–activity relationship of the 4-quinolinoyl-Gly-cyanopyrrolidine scaffold with the FAP enzyme [317, 318]. Poplawski et al. also designed and characterized more than 20 boronic acid based inhibitors for FAP [319]. Based on their studies, some of the best FAPIs, which have low inhibitory potency, were identified — two are shown in Fig. 34.

Loktev et al. and Lindner et al. conjugated DOTA to the 4-quinolinoyl-Gly-cyanopyrrolidine scaffold in order to label it with radionuclides for imaging and therapeutic purposes [320, 321]. Fifteen derivatives were developed using several approaches to conjugate DOTA to the 4-quinolinoyl-Gly-cyanopyrrolidine scaffold in different positions to improve both tumour uptake and the retention time of the radiolabelled FAPIs.

The FAPI moieties were radiolabelled with ^{68}Ga as well as ^{177}Lu (Fig. 35). The ^{68}Ga complexes formed needed purification prior to administration to patients for PET–CT imaging. This is obvious, as DOTA is not the best chelate for gallium, and larger amounts of ligand are required to achieve high complexation yields.^{177}Lu FAPI complexes were obtained with high RC yields and could be used directly for in vitro and in vivo experiments. ^{177}Lu FAPI complexes were stable when incubated in human serum for 24 h.

The first FAPIs used in humans for diagnostic PET–CT scans were ^{68}Ga–FAPI-02 and ^{68}Ga–FAPI-04. Loktev et al. used ^{68}Ga–FAPI-02 to compare its potential for tumour detection with the [^{18}F]FDG radiotracer [320]. PET–CT images obtained after the administration of both radiotracers in a patient with locally advanced lung adenocarcinoma revealed that ^{68}Ga–FAPI-02 is superior, because its accumulation is significantly higher in FAP expressing tumours than [^{18}F]FDG. A proof of concept approach, using ^{68}Ga–FAPI-04 in two patients with metastasized breast cancer, demonstrated high accumulation of the tracer in cancer metastases and low uptake in normal tissues. In addition, a dose of ^{90}Y–FAPI-04 (2.9 GBq, 78 mCi) significantly reduced the pain in a patient with metastasized breast cancer [321].

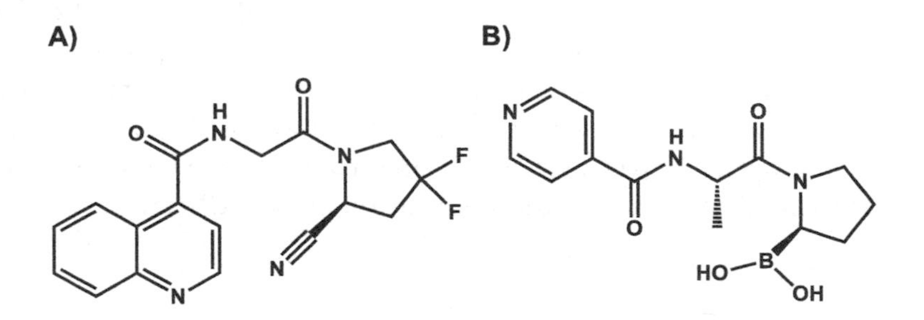

FIG. 34. FAPIs based on the 4-quinolinoyl-Gly-cyanopyrrolidine scaffold.

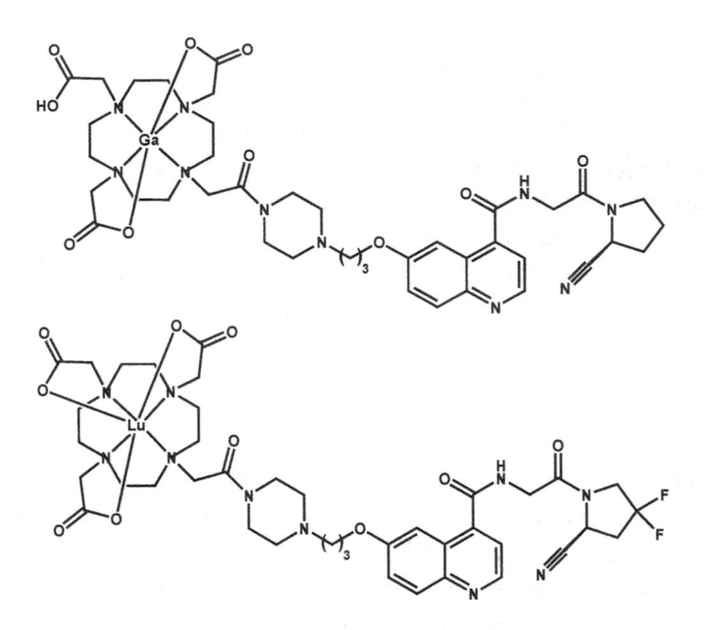

FIG. 35. ^{68}Ga and ^{177}Lu labelled FAPIs based on the 4-quinolinoyl-Gly-cyanopyrrolidine scaffold.

Giesel et al. [322] performed biodistribution and dosimetric studies for both ^{68}Ga–FAPI-02 and ^{68}Ga–FAPI-04 complexes in patients with different kinds of cancer; in particular, pancreatic, head and neck, colon, lung and breast cancer. Imaging studies demonstrated that ^{68}Ga–FAPI-02 was cleared from the tumour site more quickly than ^{68}Ga–FAPI-04 [322]. The tumour to background ratios were better in PET–CT images acquired 1 h post-injection of ^{68}Ga–FAPI-02 or ^{68}Ga–FAPI-04 than in images acquired after injecting [^{18}F]FDG.

A clinical study employing ^{68}Ga–FAPI-04 in 80 patients with 22 different types of tumour was reported by Kratochwil et al. [323]. The highest uptake was found in sarcoma, breast, cholangiocarcinoma, lung and oesophageal cancers, while differentiated thyroid, pheochromocytoma, adenoid cystic, renal cell and gastric cancer presented the lowest ^{68}Ga–FAPI-04 uptake [313]. The uptake of the tracer in muscle and the blood pool was low, which increased the tumour to background contrast ratios. Recently, ^{225}Ac–FAPI-04 was used to treat pancreatic cancer [324]. Alpha therapy that targets the fibroblast activation protein will be a new strategy for the treatment of cancers overexpressing FAP enzyme.

8.6. CONCLUSION

Sodium iodide (^{131}I) is the classic target specific radiopharmaceutical, as it is specifically transported to the follicular cells of the thyroid by an active process in which sodium iodide symporter plays an important role [325]. The iodine accumulated in the thyroid gland has a very high biological half-life, as it is stored in tyrosine residues of the thyroglobulin present in the thyroid gland. Treatment of thyroid cancer by the administration of a large activity of ^{131}I is a practice that has been in use for around 75 years. However, there was no other similar radionuclide or radiopharmaceutical that could be used for the treatment of any other form of primary cancer and bone metastases for a long time.

The development of receptor binding radiotracers as a new class of radiopharmaceuticals was suggested as early as 1979 by Eckelman et al. [326]. The localization of endocrine related tumours with a radioiodinated analogue of somatostatin was a significant development, which initiated the theranostic use of peptides as receptor targeting agents in any type of cancer [327]. As of today, ^{177}Lu–DOTA–TATE is a first line therapy for the management of neuroendocrine tumours [328]. The DOTA–TATE molecule contains the BFCA chelate DOTA, which can be used for chelating both diagnostic and therapeutic radionuclides. Peptide receptor radionuclide therapy shows great promise, as a large number of receptor molecules are overexpressed in different cancers. The development of pentixafor to target the CXCR4 receptor, which is overexpressed in the metastatic processes of more than 23 types of cancer that we know of, is another major step towards the development of theranostic radiopharmaceuticals for the management of those cancers [3]. Peptide receptor radionuclide therapy is expected to grow further in the coming years by identifying new receptor targets and their corresponding radiolabelled peptides [314].

The use of radiolabelled inhibitor molecules targeting PSMA was a milestone in radiopharmaceutical science and occurred thanks to the efforts of a large number of scientists [277]. Inhibitors are synthetically produced small molecules that resemble the substrate of the enzyme but do not undergo further metabolism after binding with the enzyme. Radiolabelled enzyme inhibitors can accumulate on the enzyme surface without further metabolism, unlike substrates. As inhibitors are small molecules, the unbound tracer is cleared through the renal system very quickly. Hence, high contrast images can be obtained when using inhibitors for diagnostic imaging. Further, large absorbed doses can be delivered to the target without affecting the healthy organs when using radiolabelled inhibitors in therapy.

The development of radiolabelled inhibitor molecules as TSRs was a great opportunity for radiopharmaceutical science and nuclear medicine. More than 75 000 enzymes are thought to exist in the human body, which are responsible for the body's metabolism. Surely many of them will be overexpressed in different diseases. It is likely that many more enzymes that are overexpressed in cancers will be identified in the future. Synthesizing enzyme inhibitors, studying their structure–activity relations and developing TSRs for cancer theranostics will be very interesting radiopharmaceutical research areas in the future.

The TSRs discussed in this section are useful not only for treating primary cancer, but also for treating metastases anywhere in the body, including in the bone. Clinical studies with large numbers of patients will be essential to identify whether such TSRs will be more advantageous than bone seeking radiopharmaceuticals for bone pain palliation.

9. FORMULATION OF BONE SEEKING RADIOPHARMACEUTICALS

The introduction of ready to use freeze-dried kits in the 1970s was a major step in diagnostic imaging using ^{99m}Tc radiopharmaceuticals [329]. The IAEA published Technical Report Series No. 466, Technetium-99m Radiopharmaceuticals: Manufacture of Kits, in 2008 [330]. This publication provided details about preparing freeze-dried kits for the formulation of 23 different ^{99m}Tc radiopharmaceuticals used for SPECT imaging [330]. The freeze-dried kits developed ought to be robust and able to provide radiopharmaceuticals with high yields, purity and safety for human administration. There is an interest in developing ready to use freeze-dried kits for making radiopharmaceuticals for bone pain palliation.

This section gives the protocol for preparing freeze-dried kits for the production of bone seeking therapeutic radiopharmaceuticals. The protocol for the formulation of therapeutic doses of ^{177}Lu–PSMA-617 and ^{225}Ac–PSMA-617 at hospital radiopharmacies is also provided at the end of the section. These protocols can be used as guidance for the formulation of these products in nuclear medicine departments.

The formulations described below have to be carried out in facilities with good laboratory practices, taking care of both radiation safety and pharmaceutical safety. Guidance on the type of facilities needed for the manufacture of kits and formulation of radiopharmaceuticals is provided in relevant IAEA technical documents [330, 331]. All operations involving radioactivity ought to be carried out with adequate shielding and safety precautions.

9.1. HEDP (ETIDRONATE)

The kit formulation given below is the same as that used for the clinical use of ^{188}Re–HEDP (etidronate), published by Mallia et al. [226].

9.1.1. Reagents

— HEDP;
— $SnCl_2 \cdot 2H_2O$;
— Ammonium perrhenate;
— Anhydrous sodium acetate;
— Gentisic acid.

9.1.2. Chemical composition of kit

— HEDP: 9 mg;
— $SnCl_2 \cdot 2H_2O$: 4 mg;
— Ammonium perrhenate: 1 mg;
— Gentisic acid: 3 mg.

9.1.3. Manufacturing formula for 22 vials

— HEDP: 198 mg;
— $SnCl_2 \cdot 2H_2O$: 88 mg;
— Gentisic acid: 66 mg;
— Final volume: 22 mL.

9.1.4. Preparation of kit solution for a batch of 20 vials of HEDP kits

— Use water for injection bubbled with nitrogen gas for all dilutions.
— Solution A: dissolve 198 mg of HEDP and 66 mg of gentisic acid in 20 mL of nitrogen purged autoclaved water.
— Solution B: dissolve 132 mg of $SnCl_2 \cdot 2H_2O$ in 500 μL of concentrated HCl by gentle heating until a clear solution is obtained. Add 2.5 mL of nitrogen purged water to achieve a final concentration of 44 mg/mL.
— Add 2 mL of solution B to solution A and mix thoroughly. The solution is passed through a sterile 0.22 μm filter.
— Dispense 1 mL of this solution into 22 autoclaved glass vials with 10 mL capacity.
— Close the vials partially with sterile slotted rubber closures. Freeze the contents of the vials. Freeze-dry.

— Seal the freeze-dried vials under vacuum and store at 2–8°C after affixing the label.
— Solution C: dissolve 27 mg (~100 μmol) of ammonium perrhenate in 10 mL of nitrogen purged autoclaved water. Pass it through a 0.22 μm filter and dispense 0.4 mL each in 22 sterile vials.
— Solution D: dissolve 2.0 g of anhydrous sodium acetate in 25 mL of nitrogen purged autoclaved water such that a final concentration of 80 mg/mL is obtained. Pass it through a 0.22 μm filter and dispense 1 mL each in 22 vials.

9.1.5. Storage of HEDP kits and Solution C and D

— 2–8°C in a refrigerator;
— Shelf life of 1 year.

9.1.6. Radiolabelling

Add 100 μL of solution C (ammonium perrhenate, 1 μmol) to 1 mL of freshly eluted $Na^{188}ReO_4$ solution (3.7 GBq, 100 mCi). Transfer it to the HEDP kit vial and keep in a boiling water bath for 20 min. Allow the vials to come to room temperature. Add 0.5 mL of solution D (sodium acetate, 40 mg) to the kit vial. Mix well to obtain the clinical dose of ^{188}Re–HEDP. Total volume is 1.5 ml.

9.1.7. Quality control

Hydrolyzed or reduced $^{188}ReO_2$ and free $^{188}ReO_4$ that are not complexed with the ligand are the two impurities that can occur during radiolabelling. Paper chromatography with Whatman No. 1 paper and two different eluting solvents — normal saline and acetone — are used for radiochemical determination of ^{188}Re–HEDP. In the saline system, ^{188}Re–HEDP and free $^{188}ReO_4$ both migrate to the solvent front and reduced $^{188}ReO_2$ remains at the point of application. By contrast, when acetone is used as the eluting solvent, ^{188}Re–HEDP and reduced $^{188}ReO_2$ remain at the point of application and free $^{188}ReO_4$ migrates to the solvent front. ^{188}Re–HEDP can be prepared to >98% radiochemical purity.

9.1.8. Shelf life of ^{188}Re–HEDP prepared using freeze-dried HEDP kits

It is recommended that the agent be administered to patients immediately after preparation. The radiochemical purity of ^{188}Re–HEDP prepared using freeze-dried HEDP kits usually remains unaltered for several hours when stored at temperatures below 30°C.

9.2. EDTMP

The protocol given below was developed for the clinical trial of ^{177}Lu–EDTMP conducted by the IAEA through CRPs [154, 246]. The kit composition is identical to that of the commercially supplied quadramet. This kit was also employed for the clinical use of ^{170}Tm–EDTMP [257].

9.2.1. Reagents

— EDTMP;
— $CaCO_3$;
— NaOH.

9.2.2. Chemical composition of kit

— EDTMP: 35 mg;
— NaOH: 14.1 mg;
— $CaCO_3$: 10.21mg.

9.2.3. Manufacturing formula for 100 EDTMP kit vials

— EDTMP: 3500 mg;
— NaOH: 1410 mg;
— $CaCO_3$: 1021 mg;
— Final volume:100 mL.

9.2.4. Preparation of kit solution for a batch of 22 vials of EDTMP kits

— Use HPLC grade water for all dilutions.
— Solution A: weigh 1021 mg $CaCO_3$ and transfer it to a 250 mL conical flask. Add 40 mL HPLC grade water to it and mix for about 5 min until it is completely dissolved.
— Add 3.5 g of EDTMP to the above solution and mix well until completely dissolved.
— Solution B: weigh 1410 mg of NaOH and transfer it to a beaker. Dissolve it in 20 mL of HPLC grade water.
— Add solution B to solution A in a dropwise manner. The solution initially becomes turbid with the addition of solution B. The turbidity disappears when the dropwise addition of solution B is continued further. The addition of solution B is continued until the pH of the reaction mixture becomes 7. Stir the resulting mixture for about 5 min.

— Add HPLC grade water to make up the volume of the solution to 100 mL.
— Filter the solution through the 022 μ Millipore filter paper.
— Dispense 1 mL aliquots of the solution into sterile glass vials, and then close them loosely with sterile rubber closures.
— Freeze the vials at −20°C overnight.
— Freeze-dry the vials under sterile conditions for 24 h. Seal the vials under vacuum.

9.2.5. Storage of EDTMP kits

— 2–8°C in a refrigerator;
— Shelf life ~1 year.

9.2.6. Formulation of patient dose of ^{153}Sm/^{170}Tm/^{177}Lu–EDTMP

Allow the freeze-dried EDTMP kits to be at ambient temperature. Add 1 mL of normal saline followed by the required volume (100–200 μL) of $^{153}SmCl_3$, $^{170}TmCl_3$ or $^{177}LuCl_3$ solution. The activity used is 1.85–3.70 GBq (50–100 mCi) for ^{153}Sm and ^{177}Lu, whereas 370 MBq (10 mCi) is used for ^{170}Tm. The reagents are incubated for 15 min at room temperature to provide ready to use formulations of ^{153}Sm–EDTMP, ^{177}Lu–EDTMP or ^{170}Tm–EDTMP suitable for human administration. All three complexes could be prepared with >98% radiochemical purity using the lyophilized EDTMP kits.

9.2.7. Quality control

Freeze-dried EDMPT kits are useful for preparing either ^{153}Sm or ^{170}Tm or ^{177}Lu–EDTMP injection. Paper chromatography with normal saline as the eluting solvent is used to estimate the radiochemical purity of these radiopharmaceutical products. ^{153}Sm/^{170}Tm/^{177}Lu–EDTMP migrate to the solvent front, showing clear demarcation with ^{153}Sm/^{170}Tm/^{177}Lu chloride, which remain at the point of application. The ^{153}Sm/^{170}Tm/^{177}Lu–EDTMP complexes can be prepared with >98% radiochemical purity using the lyophilized EDTMP kits when normal saline is used as the mobile phase with Whatman No. 1 paper.

9.2.8. Shelf life of ^{153}Sm/^{170}Tm/^{177}Lu–EDTMP prepared using freeze-dried kits

It is recommended that the agent be administered to patients immediately after preparation. The radiochemical purity of the radiopharmaceuticals prepared using freeze-dried EDTMP kits usually remains unaltered for several hours when stored at temperatures below 30°C.

9.3. DOTMP

DOTMP is a cyclic tetraamine tetraphosphonate ligand, which gives stable complexes with lanthanide radionuclides [252, 253]. The potential of ^{153}Sm–DOTMP and ^{177}Lu–DOTMP for bone pain palliation applications is well studied and limited clinical evaluation with ^{177}Lu–DOTMP has already been documented [256]. Freeze-dried DOTMP kits, suitable for the formulation of patient doses of ^{153}Sm/^{170}Tm/^{177}Lu–DOTMP, can be prepared following the methodology given below. The procedure is the same as that documented by Das et al. [252].

9.3.1. Reagents

— DOTMP;
— NaOH.

9.3.2. Chemical composition of kit

— DOTMP: 20 mg;
— NaOH: 8.75 mg.

9.3.3. Manufacturing formula for 100 DOTMP kit vials

— DOTMP: 2000 mg;
— NaOH: 875 mg;
— Final volume:100 ml.

9.3.4. Preparation of kit solution for a batch of 100 vials of DOTMP kits

— Use HPLC grade water for all dilution.
— Solution A: weigh 2 g of DOTMP and transfer it to a 250 mL conical flask. Add 40 mL HPLC grade water to it and mix for ~5 min until it is dissolved.
— Solution B: weigh 875 mg of NaOH and add to a beaker. Dissolve in 30 mL of HPLC grade water.
— Add solution B in a dropwise manner to solution A containing DOTMP until the pH of the reaction mixture reaches 7. Stir the solution for a further 5 min. Make up the volume to 100 mL using HPLC grade water.

— Filter the solution through the 0.22 μ Millipore filter paper.
— Dispense 1 mL aliquots of the solution into sterile glass vials, and then close them loosely with sterile rubber closures.
— Freeze the vials at −20°C overnight prior to freeze-drying.
— Freeze-dry the vials under sterile conditions for 24 h. Seal the vials under vacuum.

9.3.5. Storage of DOTMP kits

— 2–8°C in a refrigerator;
— Shelf life of 1 year.

9.3.6. Formulation of patient dose of $^{153}Sm/^{177}Lu$–DOTMP

Allow the freeze-dried DOTMP kits to come to ambient temperature. Add 1 mL of normal saline followed by the required volume (100–200 μL) of the $^{153}SmCl_3$ or $^{177}LuCl_3$ solution. The activity used is 1.85–3.70 GBq (50–100 mCi) for ^{153}Sm and ^{177}Lu. The reagents are incubated for 15 min at room temperature to provide ready to use formulations of ^{153}Sm–DOTMP or ^{177}Lu–DOTMP suitable for human administration. The ^{153}Sm–DOTMP or ^{177}Lu–DOTMP complexes can be prepared with >98% radiochemical purity using the lyophilized DOTMP kits.

9.3.7. Quality control

The radiochemical purity of the $^{153}Sm/^{177}Lu$–DOTMP complex prepared using the freeze-dried DOTMP kit is determined by ascending paper chromatography using normal saline as the eluting solvent, where the R_f of $^{153}Sm/^{177}Lu$–DOTMP is 0.8–1.0 and the R_f of free $^{153}Sm/^{177}Lu$ chloride is 0.0. The ^{153}Sm–DOTMP or ^{177}Lu–DOTMP complexes can be prepared with >98% radiochemical purity using the lyophilized DOTMP kits.

9.3.8. Shelf life of $^{153}Sm/^{177}Lu$–DOTMP prepared using freeze-dried kits

It is recommended that the agent be administered to patients immediately after preparation. The radiochemical purity of the radiopharmaceuticals prepared using the freeze-dried DOTMP kits usually remains unaltered for several hours when stored at temperatures below 30°C.

9.4. PREPARATION OF ^{177}LU–PSMA-617 INJECTABLE SOLUTION

Lutetium-177 labelled PSMA ligands have been used to manage castration resistant prostate cancer since their first use in 2016 [332]. A protocol for the preparation of ^{177}Lu–PSMA-617 is provided below [80]. As the other two ligands used (PSMA–I&T and iPSMA) have the same chelating molecule, DOTA, the labelling protocol described below ought to be equally applicable.

9.4.1. Reagents and materials

— PSMA-617;
— $^{177}LuCl_3$ (NCA, specific activity >3.7 GBq/μg (>100 mCi/μg));
— Ascorbic acid;
— Sodium ascorbate;
— Hydrochloric acid (30%, supra-pure);
— Ethanol (EMSURE);
— Trisodium citrate dihydrate (empura);
— Water (EMSURE);
— Acetonitrile;
— pH paper (pH 0–6);
— C18 light cartridges;
— 0.22 μm pore size syringe filters;
— 5 mL syringes (3 numbers);
— Sterile vials, 10 mL (3 numbers).

9.4.2. Preparation of reagents

9.4.2.1. PSMA-617

Dissolve 1 mg of PSMA-617 in 1 mL of EMSURE water and add aliquots of 100 μL (100 μg, 96 nmol) in 1 mL Eppendorf tubes. Freeze at −20°C. One aliquot is used for each preparation.

9.4.2.2. Ascorbic acid buffer

Dissolve 100 mg of ascorbic acid and 400 mg of sodium ascorbate in 5 mL of EMSURE water. The pH of the resultant solution is ~4.5.

9.4.2.3. Citrate buffer for quality control

Dissolve 1.47 g of trisodium citrate in 50 mL of EMSURE water; add 500 μL of 30% HCl to this solution to obtain a final pH ~5.

9.4.3. Preparation of ^{177}Lu–PSMA

— Transfer the PSMA ligand (100 μl, 100 μg) to a 10 mL autoclaved vial;
— Add 1 mL of ascorbate buffer to the above vial;
— Add $^{177}LuCl_3$ (200–400 μL) containing 925 MBq (250 mCi) to the above solution;
— Check the pH of the solution; it ought to be ~4.5–5; if needed add buffer in 0.1 mL increments;
— Heat the vial in a water bath for 30 min at 95 °C.

9.4.4. Purification

— Condition a C18 cartridge by passing 5 mL of 70% ethanol followed by 10 mL of EMSURE water.
— Condition the Millipore membrane filter (0.22 m) by passing through 2 mL of 60% ethanol. Pass 5 mL of air through using a syringe to feel the pressure to test the integrity of the filter and remove excess ethanol in it. Fit a needle to it and plunge into an autoclaved vial marked product.
— Pass the reaction mixture through the cartridge and collect the effluent in a closed vial fitted with a needle as a vent and marked 'waste'. Pass 5 mL of saline solution through the cartridge and collect the wash solution in the waste vial.
— Remove the cartridge from the waste vial and affix the syringe filter to the receiving end of the cartridge.
— Plunge a needle in the rubber closure of the vial marked 'product'.
— Plunge the needle fitted to the syringe filter and cartridge into the product vial.
— Push 1 mL of 70% ethanol through the cartridge using a 5 mL syringe.
— Push 5 mL of 0.9% saline solution to elute the product.
— Measure the activity in the product, waste and cartridge in a dose calibrator.

9.4.5. Quality control

The radiochemical purity of the ^{177}Lu–PSMA-617 injectable solution is determined either by ascending paper chromatography using Whatman 3 MM paper and a (1 : 1) mixture of acetonitrile : water as the eluting solvent, where the R_f of ^{177}Lu–PSMA is 0.8–1.0 and the R_f of free ^{177}Lu chloride is 0.0. Another system in which silica gets thin layer chromatography is used with sodium citrate buffer as the mobile phase and shows migration of free ^{177}Lu to the solvent front (R_f 0.8–1.0) and ^{177}Lu–PSMA remains at the point of application (R_f 0.0). The ^{177}Lu–PSMA-617 complex can be prepared with >98% radiochemical purity using the above described method.

9.4.6. Preparation of ^{177}Lu–PSMA-617 with CA-^{177}Lu

^{177}Lu–PSMA-617 can also be prepared with CA–^{177}Lu with a specific activity >740 MBq/mg (20 mCi/mg). The amount of peptide used will have to be adjusted such that a ligand : metal ratio of >3 : 1 is maintained to obtain a good radiolabelling yield. SPECT images of a patient treated with ^{177}Lu–PSMA-617 prepared by using CA–^{177}Lu following the protocol described above are presented in Fig. 36 [80].

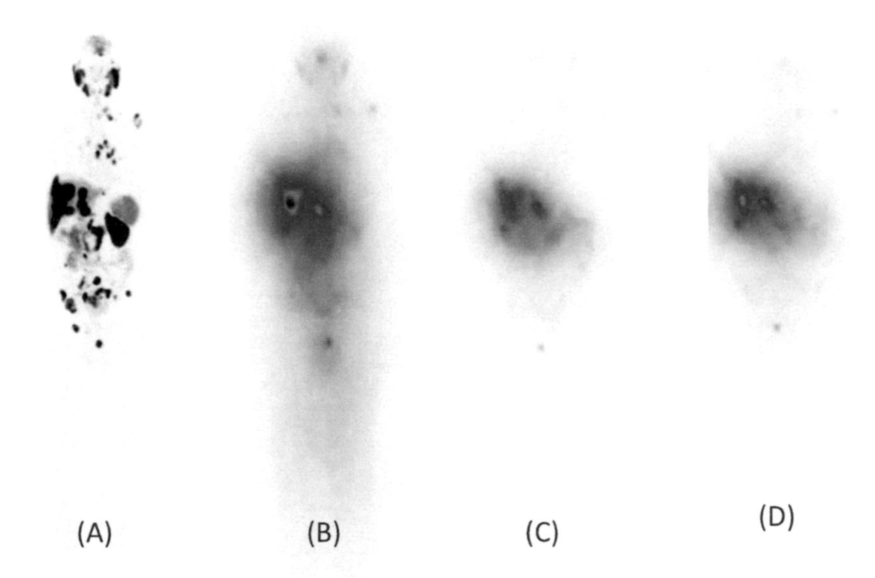

FIG. 36. [^{68}Ga]PSMA-11 PET–CT image of a patient with multiple metastases (a). Whole-body SPECT images after administration of 4.74 GBq (128 mCi) of [^{177}Lu]PSMA after 24 h (b), 72 h (c) and 7 d (d) (reproduced from Ref. [80] with permission).

9.5. PREPARATION OF ^{225}AC–PSMA-617

Alpha therapy using ^{225}Ac–PSMA-617 is very effective for the treatment of metastatic prostate cancer and is increasingly used [274]. The protocol developed for the formulation of ^{177}Lu–PSMA-617 can be adapted for the preparation of ^{225}Ac–PSMA.

Actinium-225 is delivered in 2 mL vials as a solution of 0.1 M hydrochloric acid in an approximate volume of 0.1 mL. The batch is prepared with 10 MBq (270 μCi) in order to achieve a final dose of 100 kBq/kg body weight of the patient. The reaction is performed for ~25 min and the radiochemical yield will be better than 70%. The purification and quality control are performed as described for the preparation of ^{177}Lu–PSMA-617.

Quality control tests need to be performed as quickly as possible after radiopharmaceutical preparation. This is essential because the decay products of ^{225}Ac (^{221}Fr, ^{217}At, ^{213}Bi) could be free from the chelate and hence will not move with the labelled peptide in the chromatography. The estimated RC purity could be lower than the actual purity.

REFERENCES

[1] WORLD HEALTH ORGANIZATION, Cancer, https://www.who.int/news-room/fact-sheets/detail/cancer

[2] DASH, A., KNAPP, F.F.R., PILLAI, M.R.A., Targeted radionuclide therapy — an overview, Curr. Radiopharm. **6** (2013) 152.

[3] FERRO-FLORES, G., et al., Radiolabeled protein-inhibitor peptides with rapid clinical translation towards imaging and therapy, Curr. Med. Chem. **27** (2019).

[4] INTERNATIONAL ATOMIC ENERGY AGENCY, Criteria for Palliation of Bone Metastases — Clinical Applications, IAEA-TECDOC-1549, IAEA, Vienna (2007).

[5] SILBERSTEIN, E.B., Teletherapy and radiopharmaceutical therapy of painful bone metastases, Semin. Nucl. Med. **35** (2005) 152.

[6] DASH, A., DAS, T., KNAPP, F.F.R., Targeted radionuclide therapy of painful bone metastases: Past developments, Curr. Med. Chem. **27** (2020) 3187.

[7] BAL, C. et al., Pharmacokinetic, dosimetry and toxicity study of ^{177}Lu–EDTMP in patients: Phase 0/I study, Curr. Radiopharm. **9** (2016) 71.

[8] YUAN, J. et al., Efficacy and safety of ^{177}Lu–EDTMP in bone metastatic pain palliation in breast cancer and hormone refractory prostate cancer: A phase II study, Clin. Nucl. Med. **38** (2013) 88.

[9] AGARWAL, K.K., SINGLA, S., ARORA, G., BAL, C., ^{177}Lu–EDTMP for palliation of pain from bone metastases in patients with prostate and breast cancer: A phase II study, Eur. J. Nucl. Med. Mol. Imaging **42** (2015) 79.

[10] REY, C., COMBES, C., DROUET, C., GLIMCHER, M.J., Bone mineral: Update on chemical composition and structure, Osteoporos. Int. **20** (2009) 1013.

[11] RAGGATT, L.J., PARTRIDGE, N.C., Cellular and molecular mechanisms of bone remodeling, J. Biol. Chem. **285** (2010) 25103.

[12] FERGUSON, J.L., TURNER, S.P., Bone cancer: Diagnosis and treatment principles, Am. Fam. Physician **98** (2018) 205.

[13] CLEZARDIN, P., TETI, A., Bone metastasis: Pathogenesis and therapeutic implications, Clin. Exp. Metastasis **24** (2007) 599.

[14] GUISE, T., Examining the metastatic niche: Targeting the microenvironment, Semin. Oncol. **37** (2010) S2.

[15] AVIOLI, L.V., KRANE, S. (Eds), Metabolic Bone Disease and Clinically Related Disorders, 3rd edn, Academic Press, San Diego (1997).

[16] KAKHKI, V.R.D., ANVARI, K., SADEGHI, R., MAHMOUDIAN, A., TORABIAN-KAKHKI, M., Pattern and distribution of bone metastases in common malignant tumors, Nucl. Med. Rev. Cent. East. Eur. **16** (2013) 66.

[17] RUBENS, R.D., Bone metastases — the clinical problem, Eur. J. Cancer **34** (1998) 210.

[18] SELVAGGI, G., SCAGLIOTTI, G.V., Management of bone metastases in cancer: A review, Crit. Rev. Oncol. Hematol. **56** (2005) 365.

[19] OSTA, B., BENEDETTI, G., MIOSSEC, P., Classical and paradoxical effects of TNF-α on bone homeostasis, Front. Immunol. **5** (2014) 48.

[20] AMARASEKARA, D.S., et al., Regulation of osteoclast differentiation by cytokine networks, Immune Netw. **18** (2018) e8.

[21] AHMAD, I. et al., Pain management in metastatic bone disease: A literature review, Cureus **10** (2018) e3286.

[22] MANTYH, P.W., CLOHISY, D.R., KOLTZENBURG, M., HUNT, S.P., Molecular mechanisms of cancer pain, Nat. Rev. Cancer **2** (2002) 201.

[23] ŁUKASZEWSKI, B., et al., Diagnostic methods for detection of bone metastases, Contemp. Oncol. **21** (2017) 98.

[24] BRITTON, K.E., Nuclear medicine imaging in bone metastases, Cancer Imaging **2** (2002) 84.

[25] EVEN-SAPIR, E., et al., The detection of bone metastases in patients with high-risk prostate cancer: ^{99m}Tc–MDP planar bone scintigraphy, single- and multi-field-of-view SPECT, ^{18}F-fluoride PET, and ^{18}F-fluoride PET/CT, J. Nucl. Med. **47** (2006) 287.

[26] BROOS, W.A.M., VAN DER ZANT, F.M., WONDERGEM, M., KNOL, R.J.J., Accuracy of ^{18}F–NaF PET/CT in bone metastasis detection and its effect on patient management in patients with breast carcinoma, Nucl. Med. Commun. **39** (2018) 325.

[27] POMYKALA, K.L., et al., Total-body ^{68}Ga–PSMA-11 PET/CT for bone metastasis detection in prostate cancer patients: Potential impact on bone scan guidelines, J. Nucl. Med. **61** (2020) 405.

[28] MAGEE, D.J., JHANJI, S., POULOGIANNIS, G., FARQUHAR-SMITH, P., BROWN, M.R.D., Nonsteroidal anti-inflammatory drugs and pain in cancer patients: A systematic review and reappraisal of the evidence, Br. J. Anaesth. **123** (2019) e412.

[29] LIM, F.M.Y., BOBROWSKI, A., AGARWAL, A., SILVA, M.F., Use of corticosteroids for pain control in cancer patients with bone metastases: A comprehensive literature review, Curr. Opin. Support. Palliat. Care **11** (2017) 78.

[30] WU, M.-Y., et al., Molecular regulation of bone metastasis pathogenesis, Cell. Physiol. Biochem. **46** (2018) 1423.

[31] CARDUCCI, M.A., JIMENO, A., Targeting bone metastasis in prostate cancer with endothelin receptor antagonists, Clin. Cancer Res. **12** (2006) 6296s.

[32] ZHANG, S., GANGAL, G., ULUDAĞ, H., ‘Magic bullets’ for bone diseases: Progress in rational design of bone-seeking medicinal agents, Chem. Soc. Rev. **36** (2007) 507.

[33] COSTA, L., MAJOR, P.P., Effect of bisphosphonates on pain and quality of life in patients with bone metastases, Nat. Clin. Pract. Oncol. **6** (2009) 163.

[34] GRALOW, J., TRIPATHY, D., Managing metastatic bone pain: The role of bisphosphonates, J. Pain Symptom Manage. **33** (2007) 462.

[35] KEARNS, A.E., KHOSLA, S., KOSTENUIK, P.J., Receptor activator of nuclear factor kappaB ligand and osteoprotegerin regulation of bone remodeling in health and disease, Endocr. Rev. **29** (2008) 155.

[36] STEGER, G.G., BARTSCH, R., Denosumab for the treatment of bone metastases in breast cancer: Evidence and opinion, Ther. Adv. Med. Oncol. **3** (2011) 233.

[37] CHOW, E., HARRIS, K., FAN, G., TSAO, M., SZE, W.M., Palliative radiotherapy trials for bone metastases: A systematic review, J. Clin. Oncol. **25** (2007) 1423.

[38] SERAFINI, A.N., Therapy of metastatic bone pain, J. Nucl. Med. **42** (2001) 895.

[39] PHAN, T.C.A., XU, J., ZHENG, M.H., Interaction between osteoblast and osteoclast: Impact in bone disease, Histol. Histopathol. **19** (2004) 1325.

[40] CHAKRAVARTY, R., et al., Palliative care of bone pain due to skeletal metastases: Exploring newer avenues using neutron activated ^{45}Ca, Nucl. Med. Biol. **43** (2016) 140.

[41] HOSAIN, F., SPENCER, R.P., Radiopharmaceuticals for palliation of metastatic osseous lesions: Biologic and physical background, Semin. Nucl. Med. **22** (1992) 11.

[42] LEWINGTON, V.J., Targeted radionuclide therapy for bone metastases, Eur. J. Nucl. Med. **20** (1993) 66.

[43] McEWAN, A.J.B., Use of radionuclides for the palliation of bone metastases, Semin. Radiat. Oncol. **10** (2000) 103.

[44] WISSING, M.D., VAN LEEUWEN, F.W.B., VAN DER PLUIJM, G., GELDERBLOM, H., Radium-223 chloride: Extending life in prostate cancer patients by treating bone metastases, Clin. Cancer Res. **19** (2013) 5822.

[45] CHEUNG, A., DRIEDGER, A.A., Evaluation of radioactive phosphorus in the palliation of metastatic bone lesions from carcinoma of the breast and prostate, Radiology **134** (1980) 209.

[46] SILBERSTEIN, E.B., ELGAZZAR, A.H., KAPILIVSKY, A., Phosphorus-32 radiopharmaceuticals for the treatment of painful osseous metastases, Semin. Nucl. Med. **22** (1992) 17.

[47] BARTL, R., FRISCH, B., TRESCKOW, E., BARTL, C., Bisphosphonates in Medical Practice: Actions–Side effects–Indications–Strategies, Springer, Berlin (2007).

[48] HAMDY, N.A., PAPAPOULOS, S.E., The palliative management of skeletal metastases in prostate cancer: Use of bone-seeking radionuclides and bisphosphonates, Semin. Nucl. Med. **31** (2001) 62.

[49] SUBRAMANIAN, G., McAFEE, J.G., A new complex of ^{99m}Tc for skeletal imaging, Radiology **99** (1971) 192.

[50] GOECKELER, W.F., et al., ^{153}Sm radiotherapeutic bone agents, Int. J. Radiat. Appl. Instrum. **13** (1986) 479.

[51] YANG, Y., PUSHIE, M.J., COOPER, D.M.L., DOSCHAK, M.R., Structural characterization of Sm^{III}(EDTMP), Mol. Pharm. **12** (2015) 4108.

[52] SIMON, J., et al., Organic amine phosphonic acid complexes for the treatment of calcific tumors, Google Patents (1994).

[53] PILLAI, M.R.A., Metallic Radionuclides and Therapeutic Radiopharmaceuticals, Institute of Nuclear Chemistry and Technology, Warsaw (2010).

[54] MAJALI, M.A., et al., "^{153}Sm complexes of phosphonic acid ligands", Therapeutic Applications of Radiopharmaceuticals, IAEA-TECDOC-1228, IAEA, Vienna (2001) 133–139.

[55] KOTHARI, K., et al., ^{186}Re-1,4,8,11-tetraaza cyclotetradecyl-1,4,8,11-tetramethylene phosphonic acid: A novel agent for possible use in metastatic bone-pain palliation, Nucl. Med. Biol. **28** (2001) 709.

[56] DAS, T., et al., ^{177}Lu-labeled cyclic polyaminophosphonates as potential agents for bone pain palliation, Appl. Radiat. Isot. **57** (2002) 177.

[57] CHAKRABORTY, S., et al., ^{177}Lu labelled polyaminophosphonates as potential agents for bone pain palliation, Nucl. Med. Commun. **23** (2002) 67.

[58] MATHEW, B., et al., ^{175}Yb labeled polyaminophosphonates as potential agents for bone pain palliation, Appl. Radiat. Isot. **60** (2004) 635.

[59] VATS, K., DAS, T., SARMA, H.D., BANERJEE, S., PILLAI, M.R.A., Radiolabeling, stability studies, and pharmacokinetic evaluation of thulium-170-labeled acyclic and cyclic polyaminopolyphosphonic acids, Cancer Biother. Radiopharm. **28** (2013) 737.

[60] OGAWA, K., et al., Development of [^{90}Y]DOTA-conjugated bisphosphonate for treatment of painful bone metastases, Nucl. Med. Biol. **36** (2009) 129.

[61] MECKEL, M., BERGMANN, R., MIEDERER, M., ROESCH, F., Bone targeting compounds for radiotherapy and imaging: *Me (III)–DOTA conjugates of bisphosphonic acid, pamidronic acid and zoledronic acid, EJNMMI Radiopharm. Chem. **1** (2017) 14.

[62] SMITH-JONES, P.M., et al., Radiolabeled monoclonal antibodies specific to the extracellular domain of prostate-specific membrane antigen: Preclinical studies in nude mice bearing LNCaP human prostate tumor, J. Nucl. Med. **44** (2003) 610.

[63] POMPER, M.G., et al., ^{11}C–MCG: Synthesis, uptake selectivity, and primate PET of a probe for glutamate carboxypeptidase II (NAALADase), Mol. Imaging **1** (2002) 96.

[64] DELKER, A., et al., Dosimetry for ^{177}Lu–DKFZ–PSMA-617: A new radiopharmaceutical for the treatment of metastatic prostate cancer, Eur. J. Nucl. Med. Mol. Imaging **43** (2016) 42.

[65] KRATOCHWIL, C., et al., ^{225}Ac–PSMA-617 for PSMA-targeted α-radiation therapy of metastatic castration-resistant prostate cancer, J. Nucl. Med. **57** (2016) 1941.

[66] VOLKERT, W.A., GOECKELER, W.F., EHRHARDT, G.J., KETRING, A.R., Therapeutic radionuclides: Production and decay property considerations, J. Nucl. Med. **32** (1991) 174.

[67] FINLAY, I.G., MASON, M.D., SHELLEY, M., Radioisotopes for the palliation of metastatic bone cancer: A systematic review, Lancet Oncol. **6** (2005) 392.

[68] GUERRA LIBERAL, F.D., TAVARES, A.A.S., TAVARES, J.M.R., Comparative analysis of 11 different radioisotopes for palliative treatment of bone metastases by computational methods, Med. Phys. **41** (2014) 114101.

[69] POTY, S., FRANCESCONI, L.C., McDEVITT, M.R., MORRIS, M.J., LEWIS, J.S., α-Emitters for radiotherapy: From basic radiochemistry to clinical studies — part 1, J. Nucl. Med. **59** (2018) 878.

[70] DAS, T., PILLAI, M.R.A., Options to meet the future global demand of radionuclides for radionuclide therapy, Nucl. Med. Biol. **40** (2013) 23.

[71] O'MARA, R.E., SUBRAMANIAN, G., Experimental agents for skeletal imaging, Semin. Nucl. Med. **2** (1972) 38.

[72] LANGE, R., et al., Pharmaceutical and clinical development of phosphonate-based radiopharmaceuticals for the targeted treatment of bone metastases, Bone **91** (2016) 159.

[73] THOMPSON, K., ROGERS, M.J., COXON, F.P., CROCKETT, J.C., Cytosolic entry of bisphosphonate drugs requires acidification of vesicles after fluid-phase endocytosis, Mol. Pharmacol. **69** (2006) 1624.

[74] HNATOWICH, D.J., LAYNE, W.W., CHILDS, R.L., The preparation and labeling of DTPA-coupled albumin, Int. J. Appl. Radiat. Isot. **33** (1982) 327.

[75] LIU, S., Bifunctional coupling agents for radiolabeling of biomolecules and target-specific delivery of metallic radionuclides, Adv. Drug Deliv. Rev. **60** (2008) 1347.

[76] OKOYE, N.C., BAUMEISTER, J.E., KHOSROSHAHI, F.N., HENNKENS, H.M., JURISSON, S.S., Chelators and metal complex stability for radiopharmaceutical applications, Radiochim. Acta **107** (2019) 1087.

[77] INTERNATIONAL ATOMIC ENERGY AGENCY, Technetium-99m Radiopharmaceuticals: Status and Trends, IAEA Radioisotopes and Radiopharmaceuticals Series No. 1, IAEA, Vienna (2010).

[78] CHAKRABORTY, S., et al., Preparation and biological evaluation of ^{153}Sm–DOTMP as a potential agent for bone pain palliation, Nucl. Med. Commun. **25** (2004) 1169.

[79] DAS, T., et al., ^{170}Tm–EDTMP: A potential cost-effective alternative to ^{89}SrCl(2) for bone pain palliation, Nucl. Med. Biol. **36** (2009) 561.

[80] NANABALA, R., SASIKUMAR, A., JOY, A., PILLAI, M.R.A., Preparation of [^{177}Lu] PSMA-617 using carrier added (CA) ^{177}Lu for radionuclide therapy of prostate cancer, J. Nucl. Med. Radiat. Ther. **7** (2016).

[81] WORLD HEALTH ORGANIZATION, Annex 3: Guidelines on Good Manufacturing Practices for Radiopharmaceutical Products, Technical Report Series-908, WHO, Geneva (2003).

[82] ELSINGA, P., et al., Guidance on current good radiopharmacy practice (cGRPP) for the small-scale preparation of radiopharmaceuticals, Eur. J. Nucl. Med. Mol. Imaging **37** (2010) 1049.

[83] EANM RADIOPHARMACY COMMITTEE, Guidelines on Current Good Radiopharmacy Practice (CGRPP) in the preparation of radiopharmaceuticals, EANM Radiopharmacy Committee (2007).

[84] FOOD AND DRUG ADMINISTRATION, Current Good Manufacturing Practice for Positron Emission Tomography Drugs, 21 CFR 212, US Govt Printing Office, Washington, DC (2009).

[85] MELVILLE, G., ALLEN, B.J., Cyclotron and linac production of Ac-225, Appl. Radiat. Isot. **67** (2009) 549.

[86] INTERNATIONAL ATOMIC ENERGY AGENCY, Manual for Reactor Produced Radioisotopes, IAEA-TECDOC-1340, IAEA, Vienna (2003).

[87] INTERNATIONAL ATOMIC ENERGY AGENCY, Production of Long Lived Parent Radionuclides for Generators: ^{68}Ge, ^{82}Sr, ^{90}Sr and ^{188}W, IAEA Radioisotopes and Radiopharmaceuticals Series No. 2, IAEA, Vienna (2010).

[88] KNAPP, F.F., PILLAI, M.R.A., OSSO, J.A., DASH, A., Re-emergence of the important role of radionuclide generators to provide diagnostic and therapeutic radionuclides to meet future research and clinical demands, J. Radioanal. Nucl. Chem. **302** (2014) 1053.

[89] CHAKRAVARTY, R., DASH, A., PILLAI, M.R.A., Availability of yttrium-90 from strontium-90: A nuclear medicine perspective, Cancer Biother. Radiopharm. **27** (2012) 621.

[90] PILLAI, M.R.A., DASH, A., KNAPP, F.F., Rhenium-188: Availability from the ^{188}W/^{188}Re generator and status of current applications, Curr. Radiopharm. **5** (2012) 228.

[91] INTERNATIONAL ATOMIC ENERGY AGENCY, Cyclotron Produced Radionuclides: Principles and Practice, Technical Reports Series No. 465, IAEA, Vienna (2008).

[92] INTERNATIONAL ATOMIC ENERGY AGENCY, Cyclotron Produced Radionuclides: Physical Characteristics and Production Methods, Technical Reports Series No. 468, IAEA, Vienna (2009).

[93] DASH, A., PILLAI, M.R.A., KNAPP, F.F., Production of ^{177}Lu for targeted radionuclide therapy: Available options, Nucl. Med. Mol. Imaging **49** (2015) 85.

[94] BALASUBRAMANIAN, P.S., Separation of carrier-free lutetium-177 from neutron irradiated natural ytterbium target, J. Radioanal. Nucl. Chem. **185** (1994) 305.

[95] WARE, A.R., KENDALL, F.H., Nature of phosphorus-32 produced by neutron irradiation of sulphur, Int. J. Appl. Radiat. Isot. **19** (1968) 123.

[96] SAHA, D., VITHYA, J., KUMAR, R., JOSEPH, M., Studies on purification of ^{89}Sr from irradiated yttria target by multi-column extraction chromatography using DtBuCH18-C-6/XAD-7 resin, Radiochim. Acta **107** (2019) 479.

[97] VIMALNATH, K.V., et al., Practicality of production of ^{32}P by direct neutron activation for its utilization in bone pain palliation as $Na_3[^{32}P]PO_4$, Cancer Biother. Radiopharm. **28** (2013) 423.

[98] CHUVILIN, D.Yu., et al., An interleaved approach to production of ^{99}Mo and ^{89}Sr medical radioisotopes, J. Radioanal. Nucl. Chem. **257** (2003) 59.

[99] KNAPP, F.F., DASH, A., Radiopharmaceuticals for Therapy, Springer, New Delhi (2016).

[100] RAMAMOORTHY, N., SARASWATHY, P., DAS, M.K., MEHRA, K.S., ANANTHAKRISHNAN, M., Production logistics and radionuclidic purity aspects of 153Sm for radionuclide therapy, Nucl. Med. Commun. **23** (2002) 83.

[101] BANERJEE, S., PILLAI, M.R.A., KNAPP, F.F., Lutetium-177 therapeutic radiopharmaceuticals: Linking chemistry, radiochemistry, and practical applications, Chem. Rev. **115** (2015) 2934.

[102] PILLAI, M.R.A., CHAKRABORTY, S., DAS, T., VENKATESH, M., RAMAMOORTHY, N., Production logistics of ^{177}Lu for radionuclide therapy, Appl. Radiat. Isot. **59** (2003) 109.

[103] TARASOV, V.A., et al., Production of no-carrier added lutetium-177 by irradiation of enriched ytterbium-176, Curr. Radiopharm. **8** (2015) 95.

[104] DASH, A., CHAKRAVARTY, R., KNAPP, F.F., PILLAI, M.R.A., Indirect production of no carrier added (NCA) ^{177}Lu from irradiation of enriched ^{176}Yb: Options for ytterbium/lutetium separation, Curr. Radiopharm. **8** (2015) 107.

[105] SCHWARZBACH, R., BLAEUENSTEIN, P., JEGGE, J., SCHUBIGER, P.A., Is the production of ^{186}Re with cyclotron irradiations an alternative to neutron activation in a reactor?, J. Label. Compd. Radiopharm. **37** (1995) 816.

[106] MAJALI, M.A., PILLAI, M.R.A., Preparation of $^{186/188}$Re DTPA complexes, Indian J. Nucl. Med. **8** (1993) 30.

[107] KNAPP, F.F., MIRZADEH, S., GARLAND, M.A., PONSARD, B., KUNZETSOV, R., "Reactor production and processing of ^{188}W", Production of Long Lived Parent Radionuclides for Generators: ^{68}Ge, ^{82}Sr, ^{90}Sr and ^{188}W, IAEA Radioisotopes and Radiopharmaceuticals Series No. 2, IAEA, Vienna (2010) 79–109.

[108] PONSARD, B., et al., Production of Sn-117m in the BR2 high-flux reactor, Appl. Radiat. Isot. **67** (2009) 1158.

[109] STEVENSON, N.R., et al., Methods of producing high specific activity Sn-117m with commercial cyclotrons, J. Radioanal. Nucl. Chem. **305** (2015) 99.

[110] POLYAK, A., et al., Thulium-170-labeled microparticles for local radiotherapy: Preliminary studies, Cancer Biother. Radiopharm. **29** (2014) 330.

[111] CHAKRABORTY, S., UNNI, P.R., VENKATESH, M., PILLAI, M.R.A., Feasibility study for production of ^{175}Yb: A promising therapeutic radionuclide, Appl. Radiat. Isot. **57** (2002) 295.

[112] PILLAI, M.R.A., "Ytterbium-175 production and radiopharmaceuticals", Metallic Radionuclides and Therapeutic Radiopharmaceuticals, INCT, Warsaw (2010) 152–159.

[113] CHAKRAVARTY, R., DAS, T., VENKATESH, M., DASH, A., An electro-amalgamation approach to produce ^{175}Yb suitable for radiopharmaceutical applications, Radiochim. Acta **100** (2012) 255.

[114] SALEK, N., SHIRVANI ARANI, S., BAHRAMI SAMANI, A., VOSOGHI, S., MEHRABI, M., Feasibility study for production and quality control of Yb-175 as a byproduct of no carrier added Lu-177 preparation for radiolabeling of DOTMP, Australas. Phys. Eng. Sci. Med. **41** (2018) 69.

[115] CHAKRABORTY, S., et al., Potential ^{166}Ho radiopharmaceuticals for intravascular radiation therapy (IVRT)-I: [^{166}Ho] holmium labeled ethylene dicysteine, Nucl. Med. Biol. **28** (2001) 309.

[116] BAHRAMI-SAMANI, A., et al., Production, quality control and pharmacokinetic studies of Ho–EDTMP for therapeutic applications, Sci. Pharm. **78** (2010) 423.

[117] DADACHOVA, E., MIRZADEH, S., LAMBRECHT, R.M., HETHERINGTON, E.L., KNAPP, F.F., Separation of carrier-free holmium-166 from neutron-irradiated dysprosium targets, Anal. Chem. **66** (1994) 4272.

[118] BAYOUTH, J.E., et al., Pharmacokinetics, dosimetry and toxicity of holmium-166-DOTMP for bone marrow ablation in multiple myeloma, J. Nucl. Med. **36** (1995) 730.

[119] WRIGHT, C.L., ZHANG, J., TWEEDLE, M.F., KNOPP, M.V., HALL, N.C., Theranostic imaging of yttrium-90, Biomed. Res. Int. **2015** (2015) 481279.

[120] SAMANTA, S.K., "Recovery, purification and quality evaluation of ^{90}Sr", Production of Long Lived Parent Radionuclides for Generators: ^{68}Ge, ^{82}Sr, ^{90}Sr and ^{188}W, IAEA Radioisotopes and Radiopharmaceuticals Series No. 2, IAEA, Vienna (2010) 49–78.

[121] CHAKRAVARTY, R., et al., Development of an electrochemical ^{90}Sr–^{90}Y generator for separation of ^{90}Y suitable for targeted therapy, Nucl. Med. Biol. **35** (2008) 245.

[122] PANDEY, U., DHAMI, P.S., JAGESIA, P., VENKATESH, M., PILLAI, M.R.A., Extraction paper chromatography technique for the radionuclidic purity evaluation of ^{90}Y for clinical use, Anal. Chem. **80** (2008) 801.

[123] GRACHEVA, N., et al., Production and characterization of no-carrier-added ^{161}Tb as an alternative to the clinically-applied ^{177}Lu for radionuclide therapy, EJNMMI Radiopharm. Chem. **4** (2019) 12.

[124] MÜLLER, C., et al., Terbium-161 for PSMA-targeted radionuclide therapy of prostate cancer, Eur. J. Nucl. Med. Mol. Imaging **46** (2019) 1919.

[125] CHAKRAVARTY, R., CHAKRABORTY, S., CHIRAYIL, V., DASH, A., Reactor production and electrochemical purification of ^{169}Er: A potential step forward for its utilization in in vivo therapeutic applications, Nucl. Med. Biol. **41** (2014) 163.
[126] FORMENTO-CAVAIER, R., et al., Very high specific activity erbium ^{169}Er production for potential receptor-targeted radiotherapy, Nucl. Instrum. Methods Phys. Res. Sect. B **463** (2020) 468.
[127] HENRIKSEN, G., BREISTØL, K., BRULAND, Ø.S., FODSTAD, Ø., LARSEN, R.H., Significant antitumor effect from bone-seeking, α-particle-emitting ^{223}Ra demonstrated in an experimental skeletal metastases model, Cancer Res. **62** (2002) 3120.
[128] ZALUTSKY, M.R., VAIDYANATHAN, G., Astatine-211-labeled radiotherapeutics: An emerging approach to targeted alpha-particle radiotherapy, Curr. Pharm. Des. **6** (2000) 1433.
[129] McDEVITT, M.R., et al., Radioimmunotherapy with alpha-emitting nuclides, Eur. J. Nucl. Med. **25** (1998) 1341.
[130] ROBERTSON, A.K.H., RAMOGIDA, C.F., SCHAFFER, P., RADCHENKO, V., Development of ^{225}Ac radiopharmaceuticals: TRIUMF perspectives and experiences, Curr. Radiopharm. **11** (2018) 156.
[131] GRIMM, T.L., et al., Commercial Superconducting Electron Linac for Radioisotope Production, Niowave, Inc., Lansing, MI (2015).
[132] KUKLEVA, E., KOZEMPEL, J., VLK, M., MIČOLOVÁ, P., VOPÁLKA, D., Preparation of ^{227}Ac/^{223}Ra by neutron irradiation of ^{226}Ra, J. Radioanal. Nucl. Chem. **304** (2015) 263.
[133] HENRIKSEN, G., HOFF, P., ALSTAD, J., LARSEN, R.H., ^{223}Ra for endoradiotherapeutic applications prepared from an immobilized ^{227}Ac/^{227}Th source, Radiochim. Acta **89** (2001) 661.
[134] ABOU, D.S., PICKETT, J., MATTSON, J.E., THOREK, D.L., A radium-223 microgenerator from cyclotron-produced trace actinium-227, Appl. Radiat. Isot. **119** (2017) 36.
[135] BATTAFARANO, G., ROSSI, M., MARAMPON, F., DEL FATTORE, A., Cellular and molecular mediators of bone metastatic lesions, Int. J. Mol. Sci. **19** (2018) 1709.
[136] OGAWA, K., WASHIYAMA, K., Bone target radiotracers for palliative therapy of bone metastases, Curr. Med. Chem. **19** (2012) 3290.
[137] PAES, F.M., SERAFINI, A.N., Systemic metabolic radiopharmaceutical therapy in the treatment of metastatic bone pain, Semin. Nucl. Med. **40** (2010) 89.
[138] BARRÈRE, F., VAN BLITTERSWIJK, C.A., DE GROOT, K., Bone regeneration: Molecular and cellular interactions with calcium phosphate ceramics, Int. J. Nanomedicine **1** (2006) 317.
[139] LUZ, M.A., APRIKIAN, A.G., Preventing bone complications in advanced prostate cancer, Curr. Oncol. **17** (2010) S65.
[140] CHEN, Y.-C., SOSNOSKI, D.M., MASTRO, A.M., Breast cancer metastasis to the bone: Mechanisms of bone loss, Breast Cancer Res. **12** (2010) 215.
[141] SILBERSTEIN, E.B., Systemic radiopharmaceutical therapy of painful osteoblastic metastases, Semin. Radiat. Oncol. **10** (2000) 240.

[142] CHIRBY, D., FRANCK, S., TROUTNER, D.E., Adsorption of ^{153}Sm–EDTMP on calcium hydroxyapatite, Int. J. Rad. Appl. Instrum. A **39** (1988) 495.

[143] VANDROVCOVA, M., et al., Interaction of human osteoblast-like Saos-2 and MG-63 cells with thermally oxidized surfaces of a titanium–niobium alloy, PLoS One **9** (2014) e100475.

[144] KUMAR, C., KORDE, A., KUMARI, K.V., DAS, T., SAMUEL, G., Cellular toxicity and apoptosis studies in osteocarcinoma cells, a comparison of ^{177}Lu–EDTMP and Lu–EDTMP, Curr. Radiopharm. **6** (2013) 146.

[145] GREGORY, C.A., GUNN, W.G., PEISTER, A., PROCKOP, D.J., An Alizarin red-based assay of mineralization by adherent cells in culture: Comparison with cetylpyridinium chloride extraction, Anal. Biochem. **329** (2004) 77.

[146] HALL, E.J., GIACCIA, A.J., Radiobiology for the Radiologist, 8th edn, Wolters Kluwer, Alphen aan den Rijn, Netherlands (2018).

[147] ZHAO, Q.-L., et al., Antipsychotic drugs scavenge radiation-induced hydroxyl radicals and intracellular ROS formation, and protect apoptosis in human lymphoma U937 cells, Free Radic. Res. **53** (2019) 304.

[148] DECKER, T., LOHMANN-MATTHES, M.L., A quick and simple method for the quantitation of lactate dehydrogenase release in measurements of cellular cytotoxicity and tumor necrosis factor (TNF) activity, J. Immunol. Methods **115** (1988) 61.

[149] LEGRAND, C., et al., Lactate dehydrogenase (LDH) activity of the number of dead cells in the medium of cultured eukaryotic cells as marker, J. Biotechnol. **25** (1992) 231.

[150] ALTMAN, F.P., Tetrazolium salts and formazans, J. Histochem. Cytochem. **9** (1976) 1.

[151] KUMAR, C., et al., Comparison of the efficacy of ^{177}Lu–EDTMP, ^{177}Lu–DOTMP and ^{188}Re–HEDP towards bone osteosarcoma: An in vitro study, J. Radioanal. Nucl. Chem. **319** (2019) 51.

[152] DARZYNKIEWICZ, Z., et al., Cytometry in cell necrobiology: Analysis of apoptosis and accidental cell death (necrosis), Cytometry A **27** (1997) 1.

[153] KUMAR, C., VATS, K., LOHAR, S.P., KORDE, A., SAMUEL, G., Camptothecin enhances cell death induced by ^{177}Lu–EDTMP in osteosarcoma cells, Cancer Biother. Radiopharm. **29** (2014) 317.

[154] DAS, T., SARMA, H.D., SHINTO, A., KAMALESHWARAN, K.K., BANERJEE, S., Formulation, preclinical evaluation, and preliminary clinical investigation of an in-house freeze-dried EDTMP kit suitable for the preparation of ^{177}Lu–EDTMP, Cancer Biother. Radiopharm. **29** (2014) 412.

[155] ARGUELLO, F., BAGGS, R.B., FRANTZ, C.N., A murine model of experimental metastasis to bone and bone marrow, Cancer Res. **48** (1988) 6876.

[156] PACHARINSAK, C., BEITZ, A., Animal models of cancer pain, Comp. Med. **58** (2008) 220.

[157] SCHWEI, M.J., et al., Neurochemical and cellular reorganization of the spinal cord in a murine model of bone cancer pain, J. Neurosci. **19** (1999) 10886.

[158] IANNACCONE, P.M., JACOB, H.J., Rats!, Dis. Model. Mech. **2** (2009) 206

[159] MEDHURST, S.J., et al., A rat model of bone cancer pain, Pain **96** (2002) 129.

[160] SCOTT, M.A., et al., Brief review of models of ectopic bone formation, Stem Cells Dev. **21** (2012) 655.

[161] DIXON, W.J., Efficient analysis of experimental observations, Annu. Rev. Pharmacol. Toxicol. **20** (1980) 441.

[162] INTERNATIONAL ATOMIC ENERGY AGENCY, Comparative Evaluation of Therapeutic Radiopharmaceuticals, Technical Reports Series No. 458, IAEA, Vienna (2007).

[163] MAUSNER, L.F., SRIVASTAVA, S.C., Selection of radionuclides for radioimmunotherapy, Med. Phys. **20** (1993) 503.

[164] BOUCHET, L.G., BOLCH, W.E., GODDU, S.M., HOWELL, R.W., RAO, D.V., Considerations in the selection of radiopharmaceuticals for palliation of bone pain from metastatic osseous lesions, J. Nucl. Med. **41** (2000) 682.

[165] BOUCHET, L.G., BOLCH, W.E., HOWELL, R.W., RAO, D.V., S values for radionuclides localized within the skeleton, J. Nucl. Med. **41** (2000) 189.

[166] LOEVINGER, R., BUDINGER, T., WATSON, E.E., MIRD Primer for Absorbed Dose Calculations, Society of Nuclear Medicine, Reston, VA (1991).

[167] BARDIÈS, M., CHATAL, J.F., Absorbed doses for internal radiotherapy from 22 beta-emitting radionuclides: Beta dosimetry of small spheres, Phys. Med. Biol. **39** (1994) 961.

[168] GODDU, S.M., HOWELL, R.W., BOUCHET, L.G., BOLCH, W.E., RAO, D.V., MIRD Cellular S Values: Self-Absorbed Dose per Unit Cumulated Activity for Selected Radionuclides and Monoenergetic Electron and Alpha Particle Emitters Incorporated into Different Cell Compartments, Society of Nuclear Medicine, Reston, VA (1997).

[169] COLE, A., Absorption of 20 eV to 50 000 eV electron beams in air and plastic, Radiat. Res. **38** (1969) 7.

[170] INTERNATIONAL COMMISSION ON RADIATION UNITS AND MEASUREMENTS, Stopping Powers and Ranges for Protons and Alpha Particles, ICRU Report No. 49, ICRU, Bethesda, MD (1993).

[171] VAZIRI, B., et al., MIRD Pamphlet No. 25: MIRDcell V2.0 software tool for dosimetric analysis of biologic response of multicellular populations, J. Nucl. Med. **55** (2014) 1557.

[172] GODDU, S.M., RAO, D.V., HOWELL, R.W., Multicellular dosimetry for micrometastases: Dependence of self-dose versus cross-dose to cell nuclei on type and energy of radiation and subcellular distribution of radionuclides, J. Nucl. Med. **35** (1994) 521.

[173] ARNAUD, F.X., et al., Complex cell geometry and sources distribution model for Monte Carlo single cell dosimetry with iodine 125 radioimmunotherapy, Nucl. Instrum. Methods Phys. Res. Sect. B **366** (2016) 227.

[174] MARCATILI, S., et al., Realistic multi-cellular dosimetry for ^{177}Lu-labelled antibodies: Model and application, Phys. Med. Biol. **61** (2016) 6935.

[175] HOBBS, R.F., et al., A model of cellular dosimetry for macroscopic tumors in radiopharmaceutical therapy, Med. Phys. **38** (2011) 2892.

[176] STINCHCOMB, T.G., ROESKE, J.C., Analytic microdosimetry for radioimmunotherapeutic alpha emitters, Med. Phys. **19** (1992) 1385.

[177] CHOUIN, N., BARDIÈS, M., Alpha-particle microdosimetry, Curr. Radiopharm. **4** (2011) 266.

[178] WADDINGTON, W., BARDIÈS, M., Dosimetry for Radionuclide Therapy, 104, IPEM Report, Institute of Physics and Engineering in Medicine, York (2011) 223.

[179] ROESKE, J.C., AYDOGAN, B., BARDIÈS, M., HUMM, J.L., Small-scale dosimetry: Challenges and future directions, Semin. Nucl. Med. **38** (2008) 367.

[180] BOUCHET, L.G., JOKISCH, D.W., BOLCH, W.E., A three-dimensional transport model for determining absorbed fractions of energy for electrons within trabecular bone, J. Nucl. Med. **40** (1999) 1947.

[181] ECKERMAN, K.F., STABIN, M.G., Electron absorbed fractions and dose conversion factors for marrow and bone by skeletal regions, Health Phys. **78** (2000) 199.

[182] RAJON, D.A., PATTON, P.W., SHAH, A.P., WATCHMAN, C.J., BOLCH, W.E., Surface area overestimation within three-dimensional digital images and its consequence for skeletal dosimetry, Med. Phys. **29** (2002) 682.

[183] STRIGARI, L., et al., Radiopharmaceutical therapy of bone metastases with ^{89}SrCl2, ^{186}Re–HEDP and ^{153}Sm–EDTMP: A dosimetric study using Monte Carlo simulation, Eur. J. Nucl. Med. Mol. Imaging **34** (2007) 1031.

[184] SAMARATUNGA, R.C., et al., A Monte Carlo simulation model for radiation dose to metastatic skeletal tumor from rhenium–^{186}Sn–HEDP, J. Nucl. Med. **36** (1995) 336.

[185] HOBBS, R.F., et al., A bone marrow toxicity model for ^{223}Ra α-emitter radiopharmaceutical therapy, Phys. Med. Biol. **57** (2012) 3207.

[186] SNYDER, W.S., FORD, M.R., WARNER, G.G., WATSON, S.B., 'S' Absorbed Dose per Unit Cumulated Activity for Selected Radionuclides and Organs, MIRD Pamphlet No.11, Society of Nuclear Medicine, New York (1975).

[187] XU, X.G., An exponential growth of computational phantom research in radiation protection, imaging, and radiotherapy: A review of the fifty-year history, Phys. Med. Biol. **59** (2014) R233.

[188] STABIN, M.G., SPARKS, R.B., CROWE, E., OLINDA/EXM: The second-generation personal computer software for internal dose assessment in nuclear medicine, J. Nucl. Med. **46** (2005) 1023.

[189] STABIN, M.G., SIEGEL, J.A., RADAR dose estimate report: A compendium of radiopharmaceutical dose estimates based on OLINDA/EXM version 2.0, J. Nucl. Med. **59** (2018) 154.

[190] ANDERSSON, M., JOHANSSON, L., ECKERMAN, K., MATTSSON, S., IDAC-Dose 2.1, an internal dosimetry program for diagnostic nuclear medicine based on the ICRP adult reference voxel phantoms, EJNMMI Res. **7** (2017) 88.

[191] DE JONG, M., MAINA, T., Of mice and humans: Are they the same? Implications in cancer translational research, J. Nucl. Med. **51** (2010) 501.

[192] KEENAN, M.A., STABIN, M.G., SEGARS, W.P., FERNALD, M.J., RADAR realistic animal model series for dose assessment, J. Nucl. Med. **51** (2010) 471.

[193] SEGARS, W.P., TSUI, B.M.W., FREY, E.C., JOHNSON, G.A., BERR, S.S., Development of a 4-D digital mouse phantom for molecular imaging research, Mol. Imaging Biol. **6** (2004) 149.

[194] BOUTALEB, S., et al., Impact of mouse model on preclinical dosimetry in targeted radionuclide therapy, Proc. IEEE **97** (2009) 2076.

[195] MAUXION, T., et al., Improved realism of hybrid mouse models may not be sufficient to generate reference dosimetric data, Med. Phys. **40** (2013) 052501.

[196] GARRETT, I.R., Bone destruction in cancer, Semin. Oncol. **20** (1993) 4.

[197] DeNARDO, G.L., Bone pain palliation, Cancer Biother. Radiopharm. **13** (1998) 407.

[198] PALMA, E., CORREIA, J.D.G., CAMPELLO, M.P.C., SANTOS, I., Bisphosphonates as radionuclide carriers for imaging or systemic therapy, Mol. Biosyst. **7** (2011) 2950.

[199] ATKINS, H.L., SRIVASTAVA, S.C., Radiopharmaceuticals for bone malignancy therapy, J. Nucl. Med. **26** (1998) 80.

[200] BAUMAN, G., CHARETTE, M., REID, R., SATHYA, J., Radiopharmaceuticals for the palliation of painful bone metastases — a systemic review, Radiother. Oncol. **75** (2005) 258.

[201] DAS, T., BANERJEE, S., Radiopharmaceuticals for metastatic bone pain palliation: Available options in the clinical domain and their comparisons, Clin. Exp. Metastasis **34** (2017) 1.

[202] SILBERSTEIN, E.B., The treatment of painful osseous metastases with phosphorus-32-labeled phosphates, Semin. Oncol. **20** (1993) 10.

[203] SILBERSTEIN, E.B., BUSCOMBE, J.R., McEWAN, A., TAYLOR, A.T. Jr., Society of nuclear medicine procedure guideline for palliative treatment of painful bone metastases, SNMMI **3** (2003) 147.

[204] PANDIT-TASKAR, N., BATRAKI, M., DIVGI, C.R., Radiopharmaceutical therapy for palliation of bone pain from osseous metastases, J. Nucl. Med. **45** (2004) 1358.

[205] ATKINS, H.L., Overview of nuclides for bone pain palliation, Appl. Radiat. Isot. **49** (1998) 277.

[206] SRIVASTAVA, S., DADACHOVA, E., Recent advances in radionuclide therapy, Semin. Nucl. Med. **31** (2001) 330.

[207] BLAKE, G.M., ZIVANOVIC, M.A., McEWAN, A.J., ACKERY, D.M., Sr-89 therapy: Strontium kinetics in disseminated carcinoma of the prostate, Eur. J. Nucl. Med. **12** (1986) 447.

[208] GIAMMARILE, F., MOGNETTI, T., RESCHE, I., Bone pain palliation with strontium-89 in cancer patients with bone metastases, Q. J. Nucl. Med. **45** (2001) 78.

[209] SCIUTO, R., et al., Metastatic bone pain palliation with 89-Sr and 186-Re–HEDP in breast cancer patients, Breast Cancer Res. Treat. **66** (2001) 101.

[210] BOS, S.D., An overview of current clinical experience with strontium-89 (Metastron), Prostate Suppl. **5** (1994) 23.

[211] RESCHE, I., et al., A dose-controlled study of ^{153}Sm–ethylenediaminetetramethylenephosphonate (EDTMP) in the treatment of patients with painful bone metastases, Eur. J. Cancer **33** (1997) 1583.

[212] SERAFINI, A.N., et al., Palliation of pain associated with metastatic bone cancer using samarium-153 lexidronam: A double-blind placebo-controlled clinical trial, J. Clin. Oncol. **16** (1998) 1574.

[213] SARTOR, O., et al., Samarium-153–Lexidronam complex for treatment of painful bone metastases in hormone-refractory prostate cancer, Urology **63** (2004) 940.

[214] EARY, J.F., et al., Samarium-153–EDTMP biodistribution and dosimetry estimation, J. Nucl. Med. **34** (1993) 1031.

[215] COLLINS, C., et al., Samarium-153–EDTMP in bone metastases of hormone refractory prostate carcinoma: A phase I/II trial, J. Nucl. Med. **34** (1993) 1839.

[216] LEPAREUR, N., et al., Rhenium-188 labeled radiopharmaceuticals: Current clinical applications in oncology and promising perspectives, Front. Med. **6** (2019) 132.

[217] PALMEDO, H., et al., Dose escalation study with rhenium-188 hydroxyethylidene diphosphonate in prostate cancer patients with osseous metastases, Eur. J. Nucl. Med. **27** (2000) 123.

[218] LAM, M.G.E.H., BOSMA, T.B., VAN RIJK, P.P., ZONNENBERG, B.A., ^{188}Re–HEDP combined with capecitabine in hormone-refractory prostate cancer patients with bone metastases: A phase I safety and toxicity study, Eur. J. Nucl. Med. Mol. Imaging **36** (2009) 1425.

[219] BIERSACK, H.-J., et al., Palliation and survival after repeated ^{188}Re–HEDP therapy of hormone-refractory bone metastases of prostate cancer: A retrospective analysis, J. Nucl. Med. **52** (2011) 1721.

[220] SHINTO, A.S., et al., Clinical utility of ^{188}rhenium–hydroxyethylidene-1,1–diphosphonate as a bone pain palliative in multiple malignancies, World J. Nucl. Med. **17** (2018) 228.

[221] SHARMA, R., et al., In vitro evaluation of ^{188}Re–HEDP: A mechanistic view of bone pain palliations, Cancer Biother. Radiopharm. **32** (2017) 184.

[222] CHENG, A., CHEN, S., ZHANG, Y., YIN, D., DONG, M., The tolerance and therapeutic efficacy of rhenium-188 hydroxyethylidene diphosphonate in advanced cancer patients with painful osseous metastases, Cancer Biother. Radiopharm. **26** (2011) 237.

[223] JEONG, J.M., CHUNG, J.-K., Therapy with ^{188}Re-labeled radiopharmaceuticals: An overview of promising results from initial clinical trials, Cancer Biother. Radiopharm. **18** (2003) 707.

[224] PILLAI, M.R.A., DASH, A., KNAPP, F.F., Rhenium-188: Availability from the ^{188}W/^{188}Re generator and status of current applications, Curr. Radiopharm. **5** (2012) 228.

[225] VERDERA, E.S., et al., Rhenium-188–HEDP — kit formulation and quality control, Radiochim. Acta **79** (1997) 113.

[226] MALLIA, M.B., et al., A freeze-dried kit for the preparation of ^{188}Re–HEDP for bone pain palliation: Preparation and preliminary clinical evaluation, Cancer Biother. Radiopharm. **31** (2016) 139.

[227] VAN LEEUWEN, F.W.B., VERBOOM, W., REINHOUDT, D.N., Selective extraction of naturally occurring radioactive Ra^{2+}, Chem. Soc. Rev. **34** (2005) 753.

[228] BRULAND, O.S., JONASDOTTIR, T.J., FISHER, D.R., LARSEN, R.H., Radium-223: From radiochemical development to clinical applications in targeted cancer therapy, Curr. Radiopharm. **1** (2008) 203.

[229] NILSSON, S., et al., First clinical experience with α-emitting radium-223 in the treatment of skeletal metastases, Clin. Cancer Res. **11** (2005) 4451.

[230] SHIRLEY, M., McCORMACK, P.L., Radium-223 dichloride: A review of its use in patients with castration-resistant prostate cancer with symptomatic bone metastases, Drugs **74** (2014) 579.

[231] PARKER, C., et al., Alpha emitter radium-223 and survival in metastatic prostate cancer, N. Engl. J. Med. **369** (2013) 213.
[232] KAIREMO, K., JOENSUU, T., RASULOVA, N., KILJUNEN, T., KANGASMÄKI, A., Evaluation of α-therapy with radium-223–dichloride in castration resistant metastatic prostate cancer — the role of gamma scintigraphy in dosimetry and pharmacokinetics, Diagnostics **5** (2015) 358.
[233] VAIDYANATHAN, G., ZALUTSKY, M.R., Applications of ^{211}At and ^{223}Ra in targeted α-particle radiotherapy, Curr. Radiopharm. **4** (2011) 283.
[234] ABI-GHANEM, A.S., McGRATH, M.A., JACENE, H.A., Radionuclide therapy for osseous metastases in prostate cancer, Semin. Nucl. Med. **45** (2015) 66.
[235] MAXON, H.R., et al., Rhenium-186 hydroxyethylidene diphosphonate for the treatment of painful osseous metastases, Semin. Nucl. Med. **22** (1992) 33.
[236] KOTHARI, K., et al., Preparation, stability studies and pharmacological behavior of [^{186}Re]Re–HEDP, Appl. Radiat. Isot. **51** (1999) 51.
[237] VAN DODEWAARD-DE JONG, J.M., et al., A phase I study of combined docetaxel and repeated high activity ^{186}Re–HEDP in castration-resistant prostate cancer (CRPC) metastatic to bone (the TAXIUM trial), Eur. J. Nucl. Med. Mol. Imaging **38** (2011) 1990.
[238] PIRAYESH, E., et al., Phase 2 study of a high dose of ^{186}Re–HEDP for bone pain palliation in patients with widespread skeletal metastases, J. Nucl. Med. Technol. **41** (2013) 192.
[239] SRIVASTAVA, S.C., et al., The development and in-vivo behavior of tin containing radiopharmaceuticals — I. Chemistry, preparation, and biodistribution in small animals, Int. J. Nucl. Med. Biol. **12** (1985) 167.
[240] OSTER, Z.H., et al., The development and in-vivo behavior of tin containing radiopharmaceuticals — II. Autoradiographic and scintigraphic studies in normal animals and in animal models of bone disease, Int. J. Nucl. Med. Biol. **12** (1985) 175.
[241] KRISHNAMURTHY, G.T., et al., Tin-^{117}m(4+)DTPA: Pharmacokinetics and imaging characteristics in patients with metastatic bone pain, J. Nucl. Med. **38** (1997) 230.
[242] SRIVASTAVA, S.C., et al., Treatment of metastatic bone pain with tin-^{117}m stannic diethylenetriaminepentaacetic acid: A phase I/II clinical study, Clin. Cancer Res. **4** (1998) 61.
[243] BISHAYEE, A., et al., Marrow-sparing effects of ^{117}mSn(4+)diethylenetriaminepentaacetic acid for radionuclide therapy of bone cancer, J. Nucl. Med. **41** (2000) 2043.
[244] MASLOV, O.D., et al., Production of ^{117m}Sn with high specific activity by cyclotron, Appl. Radiat. Isot. **69** (2011) 965.
[245] GUERRA LIBERAL, F.D.C., TAVARES, A.A.S., TAVARES, J.M.R.S., Palliative treatment of metastatic bone pain with radiopharmaceuticals: A perspective beyond Strontium-89 and Samarium-153, Appl. Radiat. Isot. **110** (2016) 87.
[246] MÁTHÉ, D., et al., Multispecies animal investigation on biodistribution, pharmacokinetics and toxicity of ^{177}Lu–EDTMP, a potential bone pain palliation agent, Nucl. Med. Biol. **37** (2010) 215.
[247] CHAKRABORTY, S., et al., Evaluation of ^{177}Lu–EDTMP in dogs with spontaneous tumor involving bone: Pharmacokinetics, dosimetry and therapeutic efficacy, Curr. Radiopharm. **9** (2016) 64.

[248] SHINTO, A.S., et al., ^{177}Lu–EDTMP for treatment of bone pain in patients with disseminated skeletal metastases, J. Nucl. Med. Technol. **42** (2014) 55.

[249] THAPA, P., et al., Clinical efficacy and safety comparison of ^{177}Lu–EDTMP with ^{153}Sm–EDTMP on an equidose basis in patients with painful skeletal metastases, J. Nucl. Med. **56** (2015) 1513.

[250] DAS, T., CHAKRABORTY, S., SARMA, H.D., BANERJEE, S., ^{177}Lu–DOTMP: A viable agent for palliative radiotherapy of painful bone metastasis, Radiochim. Acta **96** (2008) 55.

[251] LIU, S., EDWARDS, D.S., Bifunctional chelators for therapeutic lanthanide radiopharmaceuticals, Bioconjug. Chem. **12** (2001) 7.

[252] DAS, T., CHAKRABORTY, S., SARMA, H.D., BANERJEE, S., Formulation and evaluation of freeze-dried DOTMP kit for the preparation of clinical-scale ^{177}Lu–DOTMP and ^{153}Sm–DOTMP at the hospital radiopharmacy, Radiochim. Acta **103** (2015) 595.

[253] CHAKRABORTY, S., et al., A 'mix-and-use' approach for formulation of human clinical doses of ^{177}Lu–DOTMP at hospital radiopharmacy for management of pain arising from skeletal metastases, J. Labelled Compd. Radiopharm. **60** (2017) 410.

[254] CHAKRABORTY, S., DAS, T., SARMA, H.D., VENKATESH, M., BANERJEE, S., Comparative studies of ^{177}Lu–EDTMP and ^{177}Lu–DOTMP as potential agents for palliative radiotherapy of bone metastasis, Appl. Radiat. Isot. **66** (2008) 1196.

[255] BRYAN, J.N., et al., Comparison of systemic toxicities of ^{177}Lu–DOTMP and ^{153}Sm–EDTMP administered intravenously at equivalent skeletal doses to normal dogs, J. Nucl. Med. Technol. **37** (2009) 45.

[256] DAS, T., SHINTO, A., KARUPPUSWAMY KAMALESHWARAN, K., BANERJEE, S., Theranostic treatment of metastatic bone pain with ^{177}Lu–DOTMP, Clin. Nucl. Med. **41** (2016) 966.

[257] DAS, T., SHINTO, A., KARUPPUSWAMY KAMALESHWARAN, K., BANERJEE, S., ^{170}Tm–EDTMP: A prospective alternative of ^{89}SrCl2 for theranostic treatment of metastatic bone pain, Clin. Nucl. Med. **42** (2017) 235.

[258] HALEY, T.J., et al., Pharmacology and toxicology of terbium, thulium, and ytterbium chlorides, Toxicol. Appl. Pharmacol. **5** (1963) 427.

[259] RUSSELL, R.G.G., Bisphosphonates: From bench to bedside, Ann. N.Y. Acad. Sci. **1068** (2006) 367.

[260] MUKHERJEE, S., SONG, Y., OLDFIELD, E., NMR investigations of the static and dynamic structures of bisphosphonates on human bone: A molecular model, J. Am. Chem. Soc. **130** (2008) 1264.

[261] ROELOFS, A.J., THOMPSON, K., GORDON, S., ROGERS, M.J., Molecular mechanisms of action of bisphosphonates: Current status, Clin. Cancer Res. **12** (2006) 6222s.

[262] KAVANAGH, K.L., et al., The molecular mechanism of nitrogen-containing bisphosphonates as antiosteoporosis drugs, Proc. Natl. Acad. Sci. U.S.A. **103** (2006) 7829.

[263] BODEI, L., et al., EANM procedure guideline for treatment of refractory metastatic bone pain, Eur. J. Nucl. Med. Mol. Imaging **35** (2008) 1934.

[264] FELLNER, M., et al., ^{68}Ga–BPAMD: PET-imaging of bone metastases with a generator based positron emitter, Nucl. Med. Biol. **39** (2012) 993.

[265] FELLNER, M., et al., PET/CT imaging of osteoblastic bone metastases with ^{68}Ga–bisphosphonates: First human study, Eur. J. Nucl. Med. Mol. Imaging **37** (2010) 834.

[266] YOUSEFNIA, H., et al., Preparation and biological assessment of ^{177}Lu–BPAMD as a high potential agent for bone pain palliation therapy: Comparison with ^{177}Lu–EDTMP, J. Radioanal. Nucl. Chem. **307** (2016) 1243.

[267] PFANNKUCHEN, N., et al., Novel radiolabeled bisphosphonates for PET diagnosis and endoradiotherapy of bone metastases, Pharmaceuticals **10** (2017) 45.

[268] MECKEL, M., et al., Development of a [^{177}Lu]BPAMD labeling kit and an automated synthesis module for routine bone targeted endoradiotherapy, Cancer Biother. Radiopharm. **30** (2015) 94.

[269] RUSSELL, R.G.G., WATTS, N.B., EBETINO, F.H., ROGERS, M.J., Mechanisms of action of bisphosphonates: Similarities and differences and their potential influence on clinical efficacy, Osteoporos. Int. **19** (2008) 733.

[270] KHAWAR, A., et al., Biodistribution and post-therapy dosimetric analysis of [^{177}Lu] Lu–DOTA^{ZOL} in patients with osteoblastic metastases: First results, EJNMMI Res. **9** (2019) 102.

[271] GRAF, F., et al., DNA double strand breaks as predictor of efficacy of the alpha-particle emitter Ac-225 and the electron emitter Lu-177 for somatostatin receptor targeted radiotherapy, PLoS One **9** (2014) e88239.

[272] BAIDOO, K.E., YONG, K., BRECHBIEL, M.W., Molecular pathways: Targeted α-particle radiation therapy, Clin. Cancer Res. **19** (2013) 530.

[273] KIM, Y.-S., BRECHBIEL, M.W., An overview of targeted alpha therapy, Tumour Biol. **33** (2012) 573.

[274] KRATOCHWIL, C., et al., Targeted α-therapy of metastatic castration-resistant prostate cancer with ^{225}Ac–PSMA-617: Dosimetry estimate and empiric dose finding, J. Nucl. Med. **58** (2017) 1624.

[275] PFANNKUCHEN, N., BAUSBACHER, N., PEKTOR, S., MIEDERER, M., ROSCH, F., In vivo evaluation of [^{225}Ac]Ac–DOTA^{ZOL} for α-therapy of bone metastases, Curr. Radiopharm. **11** (2018) 223.

[276] LACOEUILLE, F., ARLICOT, N., FAIVRE-CHAUVET, A., Targeted alpha and beta radiotherapy: An overview of radiopharmaceutical and clinical aspects, Med. Nucl. **42** (2018) 32.

[277] PILLAI, M.R.A., NANABALA, R., JOY, A., SASIKUMAR, A., KNAPP, F.F., Radiolabeled enzyme inhibitors and binding agents targeting PSMA: Effective theranostic tools for imaging and therapy of prostate cancer, Nucl. Med. Biol. **43** (2016) 692.

[278] BRECHBIEL, M.W., Bifunctional chelates for metal nuclides, Q. J. Nucl. Med. Mol. Imaging **52** (2008) 166.

[279] MURPHY, G.P., ELGAMAL, A.A., SU, S.L., BOSTWICK, D.G., HOLMES, E.H., Current evaluation of the tissue localization and diagnostic utility of prostate specific membrane antigen, Cancer **83** (1998) 2259.

[280] GHOSH, A., HESTON, W.D.W., Tumor target prostate specific membrane antigen (PSMA) and its regulation in prostate cancer, J. Cell. Biochem. **91** (2004) 528.

[281] SODEE, D.B., et al., Multicenter ProstaScint imaging findings in 2154 patients with prostate cancer, Urology **56** (2000) 988.

[282] VALLABHAJOSULA, S., et al., Prediction of myelotoxicity based on bone marrow radiation-absorbed dose: Radioimmunotherapy studies using ^{90}Y- and ^{177}Lu-labeled J591 antibodies specific for prostate-specific membrane antigen, J. Nucl. Med. **46** (2005) 850.

[283] DAVIS, M.I., BENNETT, M.J., THOMAS, L.M., BJORKMAN, P.J., Crystal structure of prostate-specific membrane antigen, a tumor marker and peptidase, Proc. Natl. Acad. Sci. U.S.A. **102** (2005) 5981.

[284] BARINKA, C., et al., Structural basis of interactions between human glutamate carboxypeptidase II and its substrate analogs, J. Mol. Biol. **376** (2008) 1438.

[285] BARINKA, C., STARKOVA, J., KONVALINKA, J., LUBKOWSKI, J., A high-resolution structure of ligand-free human glutamate carboxypeptidase II, Acta Crystallogr. Sect. F: Struct. Biol. Cryst. Commun. **63** (2007) 150.

[286] CHANG, S.S., Overview of prostate-specific membrane antigen, Rev. Urol. **6** Suppl 10 (2004) S13.

[287] BAŘINKA, C., ROJAS, C., SLUSHER, B., POMPER, M., Glutamate carboxypeptidase II in diagnosis and treatment of neurologic disorders and prostate cancer, Curr. Med. Chem. **19** (2012) 856.

[288] TANEJA, S.S., ProstaScint(R) scan: Contemporary use in clinical practice, Rev. Urol. **6** Suppl 10 (2004) S19.

[289] McGUIRE, J.J., Anticancer antifolates: Current status and future directions, Curr. Pharm. Des. **9** (2003) 2593.

[290] ZHOU, J., NEALE, J.H., POMPER, M.G., KOZIKOWSKI, A.P., NAAG peptidase inhibitors and their potential for diagnosis and therapy, Nat. Rev. Drug Discov. **4** (2005) 1015.

[291] BANERJEE, S.R., et al., Synthesis and evaluation of technetium-99m- and rhenium-labeled inhibitors of the prostate-specific membrane antigen (PSMA), J. Med. Chem. **51** (2008) 4504.

[292] KULARATNE, S.A., ZHOU, Z., YANG, J., POST, C.B., LOW, P.S., Design, synthesis, and preclinical evaluation of prostate-specific membrane antigen targeted ^{99m}Tc-radioimaging agents, Mol. Pharm. **6** (2009) 790.

[293] HILLIER, S.M., et al., ^{99m}Tc-labeled small-molecule inhibitors of prostate-specific membrane antigen for molecular imaging of prostate cancer, J. Nucl. Med. **54** (2013) 1369.

[294] BANERJEE, S.R., et al., Effect of chelators on the pharmacokinetics of ^{99m}Tc-labeled imaging agents for the prostate-specific membrane antigen (PSMA), J. Med. Chem. **56** (2013) 6108.

[295] BANERJEE, S.R., et al., ^{68}Ga-labeled inhibitors of prostate-specific membrane antigen (PSMA) for imaging prostate cancer, J. Med. Chem. **53** (2010) 5333.

[296] EDER, M., et al., ^{68}Ga-complex lipophilicity and the targeting property of a urea-based PSMA inhibitor for PET imaging, Bioconjug. Chem. **23** (2012) 688.

[297] BENEŠOVÁ, M., et al., Preclinical evaluation of a tailor-made DOTA-conjugated PSMA inhibitor with optimized linker moiety for imaging and endoradiotherapy of prostate cancer, J. Nucl. Med. **56** (2015) 914.
[298] MARTELL, A.E., MOTEKAITIS, R.J., CLARKE, E.T., HARRISON, J.J., Synthesis of *N,N'*-di(2-hydroxybenzyl)ethylenediamine-*N,N'*-diacetic acid (HBED) and derivatives, Can. J. Chem. **64** (1986) 449.
[299] HECK, M.M., et al., Treatment outcome, toxicity, and predictive factors for radioligand therapy with ^{177}Lu–PSMA–I&T in metastatic castration-resistant prostate cancer, Eur. Urol. **75** (2019) 920.
[300] HOFMAN, M.S., et al., [^{177}Lu]–PSMA-617 radionuclide treatment in patients with metastatic castration-resistant prostate cancer (LuPSMA trial): A single-centre, single-arm, phase 2 study, Lancet Oncol. **19** (2018) 825.
[301] SANTOS-CUEVAS, C., et al., ^{177}Lu-DOTA-HYNIC-Lys(Nal)-Urea-Glu: Biokinetics, dosimetry, and evaluation in patients with advanced prostate cancer, Contrast Media Mol. Imaging **2018** (2018).
[302] SGOUROS, G., HOBBS, R.F., Dosimetry for radiopharmaceutical therapy, Semin. Nucl. Med. **44** (2014) 172.
[303] KRATOCHWIL, C., et al., Targeted α-therapy of metastatic castration-resistant prostate cancer with ^{225}Ac–PSMA-617: Swimmer-plot analysis suggests efficacy regarding duration of tumor control, J. Nucl. Med. **59** 5 (2018) 795.
[304] DE MEDEIROS, R.B., GRIGOLON, M.V., ARAÚJO, T.P., SROUGI, M., Metastatic castration-resistant prostate cancer (mCRPC) treated with ^{225}Ac–PSMA-617, Braz. J. Oncol. **15** (2019) 1.
[305] WÜSTEMANN, T., et al., Design of internalizing PSMA-specific Glu-ureido-based radiotherapeuticals, Theranostics **6** (2016) 1085.
[306] MORGENSTERN, A., et al., An overview of targeted alpha therapy with ^{225}actinium and ^{213}bismuth, Curr. Radiopharm. **11** (2018) 200.
[307] NILSSON, S., et al., Bone-targeted radium-223 in symptomatic, hormone-refractory prostate cancer: A randomised, multicentre, placebo-controlled phase II study, Lancet Oncol. **8** (2007) 587.
[308] HERNÁNDEZ-JIMÉNEZ, T., et al., ^{177}Lu-DOTA-HYNIC-Lys(Nal)-Urea-Glu: Synthesis and assessment of the ability to target the prostate specific membrane antigen, J. Radioanal. Nucl. Chem. **318** (2018) 2059.
[309] LAPA, C., et al., ^{68}Ga–pentixafor–PET/CT for imaging of chemokine receptor 4 expression in glioblastoma, Theranostics **6** (2016) 428.
[310] GOURNI, E., et al., PET of CXCR4 expression by a ^{68}Ga-labeled highly specific targeted contrast agent, J. Nucl. Med. **52** (2011) 1803.
[311] HERRMANN, K., et al., Biodistribution and radiation dosimetry for the chemokine receptor CXCR4-targeting probe ^{68}Ga–pentixafor, J. Nucl. Med. **56** (2015) 410.
[312] LAPA, C., et al., CXCR4-directed endoradiotherapy induces high response rates in extramedullary relapsed multiple myeloma, Theranostics **7** (2017) 1589.
[313] AKASHI, T., et al., Chemokine receptor CXCR4 expression and prognosis in patients with metastatic prostate cancer, Cancer Sci. **99** (2008) 539.

[314] DASH, A., CHAKRABORTY, S., PILLAI, M.R.A., KNAPP, F.F., Peptide receptor radionuclide therapy: An overview, Cancer Biother. Radiopharm. **30** (2015) 47.

[315] MELETTA, R., et al., Evaluation of the radiolabeled boronic acid-based FAP inhibitor MIP-1232 for atherosclerotic plaque imaging, Molecules **20** (2015) 2081.

[316] HAMSON, E.J., KEANE, F.M., THOLEN, S., SCHILLING, O., GORRELL, M.D., Understanding fibroblast activation protein (FAP): Substrates, activities, expression and targeting for cancer therapy, Proteomics. Clin. Appl. **8** (2014) 454.

[317] JANSEN, K., et al., Extended structure–activity relationship and pharmacokinetic investigation of (4-quinolinoyl)glycyl-2-cyanopyrrolidine inhibitors of fibroblast activation protein (FAP), J. Med. Chem. **57** (2014) 3053.

[318] JANSEN, K., et al., Selective inhibitors of fibroblast activation protein (FAP) with a (4-quinolinoyl)-glycyl-2-cyanopyrrolidine scaffold, ACS. Med. Chem. Lett. **4** (2013) 491.

[319] POPLAWSKI, S.E., et al., Identification of selective and potent inhibitors of fibroblast activation protein and prolyl oligopeptidase, J. Med. Chem. **56** (2013) 3467.

[320] LOKTEV, A., et al., A tumor-imaging method targeting cancer-associated fibroblasts, J. Nucl. Med. **59** (2018) 1423.

[321] LINDNER, T., et al., Development of quinoline-based theranostic ligands for the targeting of fibroblast activation protein, J. Nucl. Med. **59** (2018) 1415.

[322] GIESEL, F.L., et al., ^{68}Ga–FAPI PET/CT: Biodistribution and preliminary dosimetry estimate of 2 DOTA-containing FAP-targeting agents in patients with various cancers, J. Nucl. Med. **60** (2019) 386.

[323] KRATOCHWIL, C., et al., ^{68}Ga–FAPI PET/CT: Tracer uptake in 28 different kinds of cancer, J. Nucl. Med. **60** (2019) 801.

[324] WATABE, T., et al., Theranostics targeting fibroblast activation protein in the tumor stroma: ^{64}Cu- and ^{225}Ac-labeled FAPI-04 in pancreatic cancer xenograft mouse models, J. Nucl. Med. **61** (2020) 563.

[325] CHUNG, J.-K., Sodium iodide symporter: Its role in nuclear medicine, J. Nucl. Med. **43** (2002) 1188.

[326] ECKELMAN, W.C., et al., Receptor-binding radiotracers: A class of potential radiopharmaceuticals, J. Nucl. Med. **20** (1979) 350.

[327] KRENNING, E.P., et al., Localisation of endocrine-related tumours with radioiodinated analogue of somatostatin, Lancet **1** (1989) 242.

[328] BODEI, L., et al., The joint IAEA, EANM, and SNMMI practical guidance on peptide receptor radionuclide therapy (PRRNT) in neuroendocrine tumours, Eur. J. Nucl. Med. Mol. Imaging **40** (2013) 800.

[329] ECKELMAN, W., RICHARDS, P., Instant ^{99m}Tc–DTPA, J. Nucl. Med. **11** (1970) 761.

[330] INTERNATIONAL ATOMIC ENERGY AGENCY, Technetium-99m Radiopharmaceuticals: Manufacture of Kits, Technical Reports Series No. 466, IAEA, Vienna (2008).

[331] INTERNATIONAL ATOMIC ENERGY AGENCY, Operational Guidance on Hospital Radiopharmacy: A Safe and Effective Approach, IAEA, Vienna (2008).

[332] KRATOCHWIL, C., et al., PSMA-targeted radionuclide therapy of metastatic castration-resistant prostate cancer with 177Lu-labeled PSMA-617, J. Nucl. Med. **57** (2016) 1170.

ABBREVIATIONS

BFCA	bifunctional chelating agent
BPAMD	(4-{[(bis(phosphonomethyl))carbamoyl]methyl}-7,10-bis(carboxymethyl)-1,4,7,10-tetraaza-cyclododec-1-yl) acetic acid)
CRP	coordinated research project
CT	computed tomography
CXCR4	chemokine-4 receptor
DBMP	demineralized bone matrix particle
DNA	deoxyribonucleic acid
DOTA	1,4,710-tetraazacyclododecane-1,4,7,10-tetraacetic acid
DOTMP	1,4,7,10-tetraazacyclododecane-1,4,7,10-tetramethylene phosphonic acid
DTPA	diethylenetriaminepentaacetic acid
EBRT	external beam radiation therapy
EDTMP	ethylene diamine tetramethylene phosphonic acid
FAP	fibroblast activation protein
FAPI	FAP inhibitor
HA	hydroxyapatite
HEDP	hydroxyethylidene diphosphonic acid
HPLC	high performance liquid chromatography
LDH	lactate dehydrogenase
LET	linear energy transfer
MDP	methylene diphosphonic
NCA	no carrier added
PET	positron emission tomography
PSMA	prostate specific membrane antigen
SOP	standard operating procedure
SPECT	single photon emission computed tomography
SUV	standard uptake value
TRT	targeted radionuclide therapy
TSR	target specific radiopharmaceutical

CONTRIBUTORS TO DRAFTING AND REVIEW

Bardiès, M.	Institut de Recherche en Cancérologie de Montpellier, France
Bilewicz, A.	Institute of Nuclear Chemistry and Technology, Poland
Das, T.	Bhabha Atomic Research Centre, India
Ferro Flores, G.	Instituto Nacional de Investigaciones Nucleares, Mexico
Giammarile, F.	International Atomic Energy Agency
Jurisson, S.	University of Missouri, United States of America
Korde, A.	International Atomic Energy Agency
Kumar, C.	Bhabha Atomic Research Centre, India
Pillai, M.R.A.	Molecular Group of Companies, India

Consultants Meeting
Vienna, Austria: 19–22 November 2018

No. 26

ORDERING LOCALLY

IAEA priced publications may be purchased from the sources listed below or from major local booksellers.

Orders for unpriced publications should be made directly to the IAEA. The contact details are given at the end of this list.

NORTH AMERICA

Bernan / Rowman & Littlefield
15250 NBN Way, Blue Ridge Summit, PA 17214, USA
Telephone: +1 800 462 6420 • Fax: +1 800 338 4550
Email: orders@rowman.com • Web site: www.rowman.com/bernan

REST OF WORLD

Please contact your preferred local supplier, or our lead distributor:

Eurospan Group
Gray's Inn House
127 Clerkenwell Road
London EC1R 5DB
United Kingdom

Trade orders and enquiries:
Telephone: +44 (0)176 760 4972 • Fax: +44 (0)176 760 1640
Email: eurospan@turpin-distribution.com

Individual orders:
www.eurospanbookstore.com/iaea

For further information:
Telephone: +44 (0)207 240 0856 • Fax: +44 (0)207 379 0609
Email: info@eurospangroup.com • Web site: www.eurospangroup.com

Orders for both priced and unpriced publications may be addressed directly to:

Marketing and Sales Unit
International Atomic Energy Agency
Vienna International Centre, PO Box 100, 1400 Vienna, Austria
Telephone: +43 1 2600 22529 or 22530 • Fax: +43 1 26007 22529
Email: sales.publications@iaea.org • Web site: www.iaea.org/publications

22-05628E-T